SLASHBURNER

Nick Raeside

SLASHBURNER

HOT TIMES IN THE BRITISH COLUMBIA WOODS

HARBOUR
PUBLISHING

COPYRIGHT © 2020 NICK RAESIDE
1 2 3 4 5 — 24 23 22 21 20

All rights reserved. No part of this publication may be reproduced, stored in a retrieval system or transmitted, in any form or by any means, without prior permission of the publisher or, in the case of photocopying or other reprographic copying, a licence from Access Copyright, www.accesscopyright.ca, 1-800-893-5777, info@accesscopyright.ca.

HARBOUR PUBLISHING CO. LTD.
P.O. Box 219, Madeira Park, BC, VON 2H0
www.harbourpublishing.com

All photos by Nick Raeside
Map by Adrian Raeside
Edited by Pam Robertson
Cover and text design by Shed Simas/Onça Design
Printed and bound in Canada

Harbour Publishing acknowledges the support of the Canada Council for the Arts, the Government of Canada, and the Province of British Columbia through the BC Arts Council.

LIBRARY AND ARCHIVES CANADA CATALOGUING IN PUBLICATION
Title: Slashburner : hot times in the British Columbia woods / Nick Raeside.
Names: Raeside, Nick, 1952- author.
Identifiers: Canadiana (print) 20200168673 | Canadiana (ebook) 2020016886x
    | ISBN 9781550178982 (softcover) | ISBN 9781550178999 (HTML)
Subjects: LCSH: Raeside, Nick, 1952- | LCSH: Fire fighters—British Columbia—
    Biography. | LCSH: Slashburning—British Columbia—Biography. | LCSH:
    Logging—British Columbia—Biography. | LCGFT: Autobiographies.
Classification: LCC SD544.3 .R34 2020 | DDC 363.37092—dc23

To Prometheus

LOCATION MAP OF GOLDEN AND AREA

CONTENTS

Preface ix

1. Learning About Fire — 1
2. Fighting Fire — 15
3. Making More Smoke — 32
4. Fire Misbehaviour — 53
5. Fuelling the Fire — 73
6. Driving Water Uphill — 90
7. Tracks in the Mud — 113
8. Explosives on the Fireline — 132
9. Miscommunications — 158
10. Between Fires — 175
11. Pile Burning — 195

Epilogue — 213
Glossary — 216
Acknowledgements — 226
About the Author — 228

PREFACE

In North America, fire was used by early settlers as a cheap and effective way of clearing land. Sometimes it would prove to be a little too effective, as was the case in Vancouver in 1886, when a fire set for this purpose got away and burned down much of the town. The BC Forest Service, created in 1912, was responsible for ensuring that such unfortunate events weren't repeated. But fires were still used for clearing purposes. It had become obvious that accumulations of woody debris left behind after logging operations posed a significant fire risk. The first recorded deliberate burning of such material in order to reduce the hazard took place in 1913.

By the beginning of the twentieth century, the steam donkey engine made industrial logging on a large scale possible. There were some unfortunate side effects, though: sparks produced by these engines would cause accidental fires in the woods during the summer months. This happened in 1938 on Vancouver Island, resulting in the Bloedel Fire, which blackened nearly seventy-five thousand acres. A major factor contributing to this fire's size and severity was the thousands of acres of slash that had accumulated after years of logging. As a direct consequence of the fire, Section 113A of the BC Forest Act came into effect the same year. This legislation made it obligatory for logging operators to dispose of the logging slash they created by burning it in the fall, to prevent accumulations from becoming a summer fire hazard. This law was directed at operations taking place on the BC coast and Vancouver Island, and it wasn't until 1967 that Section 116 of the above act extended this requirement to the rest of the province.

Logging practices were changing by the mid-twentieth century, and better utilization of the forest crop left less flammable waste. It was also becoming evident that burning this waste could sometimes be detrimental to the forest's ability to regenerate if the fire was so hot that the soil underneath was cooked. Fire started to be used rather more scientifically as a silvicultural tool; however, the burning of logging slash still involved an element of risk, as the unfortunate residents of a small community near Salmon Arm discovered in 1973 when an escaped forestry burn ran through their properties.

Wildfires that take place during the summer get a lot of media coverage, particularly when property is lost, which has unfortunately been the case a number of times in the last two decades. The majority of the public may be unaware, however, that, at least in prior years, once the upheaval of the summer wildfire season had ended, a program of deliberate fire lighting for forest management purposes would begin. Even as smoke from wildfires still lingered in mountain valleys in various parts of the province, some fire bosses became burn bosses and prepared to generate a lot more smoke. And while many firefighters greeted the arrival of cooler weather with relief as they looked forward to well-earned days off, there were others who couldn't wait for the fall slashburning program to begin. Possibly it was the prospect of a few more weeks of employment that appealed to many, but there were a few who rather enjoyed the irony of being paid to set fires in the woods.

There have been some exceptionally good fire bosses in the BC forest fire control business, and I've been privileged to have met one or two of them. There have also been some who were not so good, and possibly there are certain individuals who would place me in the latter category. I'd put myself somewhere in between, all things considered. I probably would have taken the prize for the worst-dressed fire boss if there'd been such an award, however, and I'm fairly certain that I'm the only one to have shown up on the fireline barefoot.

At least I can say that nobody was killed or seriously injured on any of the fires and slashburns I was responsible for. In hindsight I have to admit that was possibly due to good luck as much as good management, as things did get a bit weird occasionally. Safety on the fireline was at the front of my mind most of the time, but it did get pushed aside now and then by an irresistible urge to liven things up a bit. I've always believed there's room in every workplace for a few practical jokes, even though some of my former employers haven't always agreed.

The events in this book took place exactly as described. It's been a long time since I last saw the areas we burned, but I understand that the forest is growing back quite nicely.

1
LEARNING ABOUT FIRE

My introduction to wildfire took place in New Zealand when I was seven years old. A bush fire had started near the holiday cabin we were staying in, and I joined the effort to extinguish it, using a wet burlap sack to beat out the flames. It might have been easier if I hadn't been barefoot, as the fire was burning in gorse, which is a viciously spiny plant. About ten years later I became a volunteer bush firefighter. This time I had footwear when I was out on the hillside, although it wasn't entirely satisfactory, as they were plastic sandals. Most of the time it wasn't a problem—my feet were pretty tough—but on one occasion the fire unexpectedly flared up and cut off the trail I'd planned to use as an escape route. I ended up going out through the burn and found the ground was hot enough to melt my sandals. As I was hopping around like a cat on a hot tin roof, I tried to figure out why the fire had done what it did, which was the start of my interest in wildfire behaviour.

I enjoyed firefighting, so much so that whenever a wildfire started on the hills behind the community we lived in, I'd take a few days off school and help put it out. I soon discovered that I could travel to the fire for free on the local bus if I stood at the side of the road and held out my shovel. There was no pay, no worker's compensation if you got hurt, and you'd have to supply your own lunch. The only bonus came one morning after we'd fought a bush fire all night in order to save the local hotel, along with nearby houses, and the grateful proprietor brought out several crates of beer for the thirsty volunteers. The legal drinking age in the country at that time was twenty-one and I was a few years short, but I was so grubby and soot-covered that nobody would have recognized me anyway. It had been a rough night, as the fire had chased us back down the hillside once or twice.

The tool of choice for cutting fire breaks in manuka scrub and gorse was a slasher, which is a heavy machete attached to a long wooden handle—and a rather dangerous implement if handled carelessly. Some of the access trails we were using to get to the fire's edge were actually sheep trails that had been pushed through the scrub. They were more like tunnels, and often you couldn't stand upright, in which case we'd end up crawling in a line and cursing all the gorse prickles that our bare knees were picking up along the way, as most of us were wearing shorts.

At about two a.m. on the fire that threatened the hotel, we were painfully working our way uphill in the dark when it suddenly began to get very light, and the man at the head of the crew shouted that the fire was bearing down on us rapidly. We had to get turned around fast, which wasn't the easiest thing to do in such cramped conditions, particularly when each of us was carrying a slasher. Crawling downhill is actually worse than going up for some reason, and the fire gained on us rapidly before we got to where we could stand up and make better time.

It was on one of these bush fires that I first encountered a Wajax portable fire pump, which had been set up in a creek. As I listened to the curses of the man trying to start it, I had no idea at the time that I'd end up knowing this pump intimately in the future and cursing it in my own words.

I also never imagined that I'd someday be paid to use fire as a tool, though I had found out how useful fire could be when I was in high school, the day our mathematics classroom caught fire. It was in a portable building that had a heating stove with an apparently defective chimney. I was in another part of the school when the fire alarm went off, and once we discovered what was on fire, a couple of my classmates and I went off to enjoy the sight of the classroom burning instead of heading to the required assembly location. To make quite sure it was a thorough job, we moved some bike racks in order to block access for the fire engine we could hear approaching. The school authorities weren't very happy, and they read us the riot act: impeding the firefighters, endangering lives and so on.

After leaving high school prematurely, I got a job at a petrochemical plant, where I learned a great deal about flammable liquids, as I was working in the solvent plant, where explosive fumes were a constant hazard. All electrical switches were designed so they wouldn't cause sparks, as were the telephones. We weren't allowed to wear nylon clothing due to the danger of static electricity, and naturally, smoking was strictly forbidden. Not everyone took this last rule too seriously, however, judging by the number of crushed cigarette butts that accumulated out behind the building.

It was at this plant that I obtained the ingredients to make my first batch of napalm, which I prepared over the kitchen stove in the rented house we were living in at the time. It was a nice green colour, as I'd used aviation gasoline, which has a very high octane rating, and it made quite a mess of the wall when a flaming wad of it shot out the end of the test tube. Fortunately the house was scheduled to be demolished to make way for a highway bypass.

Soon after that we moved to Canada, and I spent the next few years working as a chainman for a couple of land surveying companies on the West Coast, with a spell as an oiler for BC Ferries in between. In those days survey measurements were done with a two-hundred-foot steel tape, so a chainman would spend much of his time cutting line through the bush in order to clear a path for it. This could be hard work when a large tree happened to be in the way, but it gave me lots of practice using an axe. I eventually decided that my employment future lay elsewhere—my surveyor employers were quick to agree—and accordingly signed on with a company that provided contract forest fire control services.

•

The now defunct International Forest Fire Systems company (IFFS) had developed a method of deploying crews onto fires burning on inaccessible terrain in BC by having them rappel from helicopters. The system is still used today by the BC Wildfire Service, who took it over and named it Rapattack, dressing the crews with matching outfits in the process. When I was with IFFS we were referred to as Sky Spiders and wore any old clothes—not that I'm jealous in any way, as I dislike wearing uniforms. These days it seems that how you look on the way to a fire is as important as how effectively you put it out.

The IFFS training camp in Revelstoke, which was above the KOA campground, consisted of a couple of old construction camp buildings left over from the Mica Dam project. I was the second trainee to turn up at the otherwise empty base, which meant that we had the pick of the rooms, such as they were. The first thing I did was search through the bunkhouse, looking for light bulbs that were still intact. When I finally got light in my room, it shone on a floor that was covered in sand and cigarette butts, all of which I transferred to the adjacent room. I noticed that on the ceiling there was a large patch of what looked like fungus, which was peeling off in lumps and dropping onto the mattress. From an earlier stay at a logging camp on the coast, I knew that it's never advisable to inspect mattresses too closely. Maybe that's why the light bulbs had been removed. I pried off the worst of the fungus and threw that into the next room as well.

The cookhouse looked like it had been a playground for mice and other small creatures during the winter. I was rather curious as to why there was a half-full can of automatic transmission fluid on the shelf above the stove.

Along with the other recruits, I received ten days of training in this well-appointed camp in the spring of 1977, during which time I learned some useful skills, including how to assemble a functioning chainsaw from a heap of broken saws and worn-out parts. I also had to complete the fitness tests in order to be certified as a rappel crewman. The most interesting part of training was going up in a helicopter for the first time for live rappel training, particularly since it was a one-way flight. The only way to get back to the ground was to slide down the two-hundred-foot rope attached to the side of the aircraft as it hovered above the airport runway.

Most of us didn't enjoy the mandatory twice-daily training runs and therefore didn't push ourselves as hard as we were supposed to, although we were warned that we'd have to complete the run within the time allowed in order to qualify as a crewman. The running track went right beside a barking dog that was getting increasingly annoyed by the motley group of panting humans that would pass his territory several times a day. Finally one afternoon he'd had enough, and he rushed out to bite the last man in the ass. We ended up taking him (the man, not the dog) to the hospital to provide a source of amusement for the emergency room staff as he lay face-down on the table with his pants down, getting the teeth marks examined. It wasn't a serious wound, and nothing was done about the dog: after all, he lived there permanently, and we were just transients. The incident did have a positive effect, however, as we sped up considerably on that stretch of the run, which improved our overall time.

As part of our training we were given instruction on how to fall trees with a chainsaw, something I'd never done before, as all my previous tree felling had been done with an axe. Our training took place on a mountainside where an old forest fire had created numerous hemlock snags. It was an interesting learning experience, particularly when the instructor was demonstrating a method of making an undercut in a large snag when you can't see around to the other side. Something went slightly wrong: the snag started to fall, and he was suddenly left holding the broken-off handle of the saw while the rest of it went bouncing downhill along with all the rocks that had been dislodged by the impact of the snag. We retrieved the remains and took them back to base to add to all the other wrecked chainsaws waiting in the parts-donor heap.

Once all training was complete, I was assigned to a crew that was going to be working for Evans Forest Products in Golden. When there weren't any fires to fight, we were kept out of mischief with other work, such as roadside clearing, timber cruising and cutblock traversing. My inclusion on this crew had a lot to do with my surveying experience, although the only skill I was able to employ was my ability to use a Silva compass. I was good with this type of handheld compass—sometimes a little too good, according to whoever was following behind me at the end of the plastic measuring tape, since I'd never deviate off the line. If the compass bearing went through a swamp, I'd follow it exactly and expect my partner to wade in my wake. I think eventually I was assigned to whoever drew the short straw when we were paired up for timber cruising.

The first forest fire we were dispatched to was a small one that had been started by lightning on a steep mountainside above Beavermouth, near the east end of Glacier National Park. A helicopter flew us up in the evening to take over from the BC Forest Service suppression crew that had been fighting it—apparently they didn't want to fight fire in the dark. The other two on my crew had never been on a wildfire before, so I was elected fire boss. We started things off by carefully falling a burning snag so that it landed in the burn, then cut up all the burning logs and scraped the hot embers out into a "boneyard" (cleared dirt area) since we had no water to extinguish them. When it got too dark to see what we were doing, we used the battery-powered headlamps on our hard hats and dug fireguard until 1:30 a.m. It was then that we discovered our fire equipment pack contained only two sleeping bags. None of us relished the thought of sharing a bag, so I curled up in the dirt beside the fire. Eventually it got cold, so I went into the burn and carried on scraping off glowing embers as a way of staying warm.

It soon became obvious that much of the IFFS equipment was lacking in quality as well as quantity. The Stihl chainsaws we'd be using were worn out, and the crewcab pickup truck that had been assigned to us wasn't much better. This was made clear to us the day that one of the front wheels fell off while we were out in the woods. The problem was easily fixed, once we'd searched the gravel road for the nut and washers, but getting replacement parts for the saws was more difficult. It turned out the local Stihl agent had not been paid by IFFS the previous fire season and was therefore extremely reluctant to extend further credit. He eventually did so, grudgingly, probably because he got tired of us grovelling at the parts counter every few days.

The local supermarket cut off our credit after a few weeks too, for the same reason. IFFS did not have a good reputation for settling accounts in town. This reluctance to part with money extended to crew wages, as we discovered when we attempted to cash our first advance cheques. There wasn't enough money in their bank account to cover all of the wages, so only the first three crewmen through the door got cash. As far as I can recall, I wasn't one of them. To top it all off, the owner of the motel we were staying at hadn't been paid and was considering legal action. To keep him on our side and to prevent him from cutting off the electricity to the cabins we were occupying, a couple of us tried to make ourselves useful by helping with repairs around the property.

It wasn't a good state of affairs, as we explained to the IFFS company president when he landed briefly at the town airport on a fly-in visit. He airily dismissed our list of complaints with the prediction that it would be a lucrative fire season and the promise that the bank account would be refilled in the meantime. In retrospect, the reflective sunglasses he wore were possibly intended to prevent us from telling by his eyes that he was lying through his teeth. We noticed that he didn't show any interest in sorting out our credit problems with the local businesses before he flew off into the sunset.

Things improved somewhat when we were sent out on a timber cruising project near Tsar Creek, north of Golden, that required us to stay in a tent camp (known as a "fly camp," probably after the mosquitoes that greeted our arrival). There was a full-time cook in camp, which meant we were finally eating real food, and lots of it. Earlier that season I'd been rappelled in on my own to cut a helispot and build a landing pad in preparation for this timber cruise. Sending me in on my own in this situation was actually against the safety rules, since once I was on the ground, there was no way to extract me if I had an accident before I got the site cleared and the landing pad constructed. For some reason the crew was one man short that day, so I didn't get a partner, but I wasn't particularly concerned. I worked hard for a few hours felling trees and clearing brush before another crewman rappelled in to help drag logs into place to construct the cribbing for the pad, and we barely finished by the time the helicopter came to pick us up before daylight faded.

A trail bike was flown in along with all the other fly camp gear, so two of us could drive back and forth along the few miles of the old Big Bend Highway that hadn't been inundated by the Mica Dam. The motorbike saved a fair bit of walking, as we could ride to the start of our hike up the mountainside each morning and ride back in relative comfort at the end of the day. It worked well until my passenger on the back inadvertently allowed the axe he was holding to rub against the back tire. There was a sudden bang and we ended up pushing the motorbike the remaining distance back to camp.

That wasn't the only catastrophe during this project, as one night the rain swelled the creek next to our camp and washed our beer supply down into the lake. Most of it was subsequently recovered, but there may still be some lying in the mud at the bottom of Kinbasket Lake.

After the cruising was finished and we were back in town, we eventually got tired of waiting at the motel after work for a fire call and discovered it was more comfortable sitting in one of the local pubs. They understandably didn't extend credit, so we had to watch our diminishing cash reserves. At least draft beer was cheap back then, even if it did taste like it had passed through a horse's kidneys. While sitting at our usual table one night, we got a call-out to a fire that had been started by lightning. The location wasn't exact, and we ended up driving up and down a logging road with our heads out the windows—there was a strong smell of smoke in the air but no sign of flames in the dark. During this time we somehow managed to drive into the ditch and get thoroughly stuck, so after a group discussion we decided to bed down for the night and sort things out at first light. Two of the crew slept in the crewcab and the rest of us in the ditch, which fortunately was bone dry. When the sun rose, we discovered that by pure chance we'd stopped right next to the fire, which was smouldering in heavy timber less than a hundred feet from the truck.

•

Our IFFS crew continued to be assigned other tasks to keep us busy when we weren't fighting fires or timber cruising: roadside clearing and road sign replacement, for example. Somehow I managed to find ways to make these mundane tasks a bit more interesting. While out clearing overhanging trees from the sides of the logging roads, I'd pack a two-by-ten plank up the cutbanks, along with my chainsaw. Once I'd cut down the offending trees, I'd let the saw slide back down to the ditch, then surf down the steep gravel slope on the plank. Sometimes when I hit the bottom, I'd end up rolling out onto the road, which bothered the rest of the crew to the point that they eventually confiscated my surfboard and cut it up into pieces. I don't know why they were so concerned, as the chances of me rolling under the wheels of a passing logging truck were minimal.

This was about the time that the call-point mile markers on all the company logging roads were being changed to kilometres, thanks to the federal government's idiotic belief in the benefits of the metric system. We were sent out to replace old signs with new ones, which was rather boring until we decided to replace the markers on one active haul road without informing the logging truck drivers. They didn't know what was going on at first when they found the numbering system mysteriously changing as we drove along the road, pulling out old signs and hammering in new ones just ahead of their trucks. Fortunately there were no collisions as a result of our innocent prank, but I'm quite sure a few of the drivers would have liked to have hammered on us for a bit.

All in all, the first fire season wasn't too rewarding, as we didn't get to go out on that many fires. The ones we did work on were fairly small and were out all too soon. IFFS had a rule that a crew on a fire could declare it out and radio for helicopter pickup once they were certain the fire was completely extinguished, but should the fire flare up again after you left, you'd be sent back to put it out. Without pay this time, which was a sobering prospect. As far as I'm aware, no IFFS crew ever committed this error.

Things were better when I went back the following year, even though by this time IFFS was going through financial convulsions and in fact still owed most of us our final paycheques from the previous season. (It's been forty years now and I still haven't received mine, so I'm not holding my breath.) We were contracted once again to Evans Forest Products, and there was a better system in place to make sure we got paid properly. IFFS was still nominally our employer and was supplying the rappel equipment, but they were doing this through one of their numerous shell companies.

•

"Smoke if you wish, but be prepared to leave by the hole that appears in the roof."
—Sign attached to the wall of the flammable-solvents plant

2
FIGHTING FIRE

The 1978 fire season started on a promising note when a fire that had been burning underground over the winter spread into the nearby forest early in June. This type of fire was known as a landing hangover, and it took a bit of a run while fireguards were being constructed. We ended up being chased out when fire blew up in a gully below us. I was riding as swamper on one of the two bulldozers that were leaving in a hurry, although I probably would have made faster time on foot. Swampers were assigned to each bulldozer as operator's assistants, and we weren't supposed to ride on the machines, but it was frequently safer to do so on active forest fires. As we clanked and rattled our way out to safety, I looked back in time to see a spruce tree close by go up in flames, as they often do in forest fires. This one was unusual, though, in that a crimson flame wrapped around it in a spiral, looking much like the cellophane wrapping that's twisted around presentation bunches of flowers. It would have made a pretty picture if I'd had a camera with me.

We worked until long after nightfall on this fire, and I ended up helping one of the fallers cut down a snag by shining a flashlight for him so he could watch for the back cut to open up as an indication the snag was starting to fall. It was pitch dark all around, and we hoped there were no other firefighters nearby who might be hit when it crashed to the ground. Fortunately there was nobody unaccounted for when everyone finally assembled to pack it in for the night.

The remainder of the fire season didn't meet expectations, which was probably good for the health of the forest but not so good for our bank balances, due to the lack of overtime. We participated in timber cruising operations in some of the river valleys tributary to Kinbasket Lake, starting with the Sullivan, followed by the Kinbasket and Cummins Rivers. We did, however, rappel into two small lightning-strike fires at the back end of Bush River that turned out to be interesting due to a few things that went wrong.

A problem occurred on the first fire when the spotter sitting in the front seat of the helicopter forgot to attach me to the rope I'd be sliding down. I discovered this minor omission as I was sitting on the helicopter skid with my feet dangling over a two-hundred-foot drop. The spotter had given me the signal to descend, and I was about to let go of the bracket anchoring the rope to the helicopter when I realized I'd end up dropping in free fall. I quickly connected myself and descended, composing my speech to the spotter on the way to the ground. Once I got my hands on our portable radio, I recited the highlights of it to the spotter, who was hovering overhead. (I went over it in greater detail when I met up with him in town later.)

After we'd finished building the fireguard and putting out the worst of the hot spots, we stopped for lunch. Typical firepack food was canned beef hash, canned peaches and Stoned Wheat Thins. Baker's Semi-Sweet Chocolate might be present as an additional treat, unless some evil bastard had earlier rifled the pack. We discovered that the supply of drinking water in the firepack was putrid, as it hadn't been changed since the previous year. We ended up dumping it on the fire.

Later that day the Forest Service took pity on our lack of drinking water, and one of the assistant rangers flew over and tossed out a case of cheap canned cola soft drink as the helicopter was hovering above us. We thought he was a hero, until we discovered it was lukewarm, and when you drink the vile stuff at that temperature, it just makes you thirstier.

We were rappelled in to the second fire the following day, and a problem quickly showed up when we started to cut out a helispot. (As a safety measure on our rappel-access fires, we would clear a landing area so that it would be possible for a helicopter to come in if one of us should get seriously injured. Extinguishing the fire was a secondary priority, unless it was in imminent danger of getting away.) We discovered that for some reason the chainsaw wouldn't run, and after giving up on it in disgust, we ended up felling trees for the clearing with the two Pulaskis that were in the fire equipment pack (actually an oversized cardboard box held together with rope). A Pulaski is a combination axe and hoe tool, and both of those in the firepack were blunt, which made progress slow as we beat on the spruce trees and cursed whoever had put the firepack together for forgetting to include a file. Our day didn't improve when I discovered the reason the saw wouldn't run was that some idiot had replaced the fuel line filter with a foam earplug, which had disintegrated and plugged up the carburetor jets.

•

The following year made up for everything, however, as there were three sizeable fires, one after another. The first one started in the Smith Creek valley, northwest of Golden, when a cedar tree that was felled rubbed against another tree on its way down and the friction started a fire. The faller was unable to put it out with the small extinguisher he carried with him, so the fire took off up the mountainside. It was still going strong when I arrived on site in time to watch the air tankers drop a few loads of fire retardant. Not all of it was effective, as the valley was narrow, which made accurate flying over the fire difficult and hazardous.

Once the air show was over, it was time for the ground crews to go back to work on the fire. There were a lot of large hollow cedar trees burning within the fire perimeter, so it was risky in some places. Unfortunately the forest officer who was on site didn't seem to appreciate this fact, and he insisted that fallers go in and drop the burning cedars. It was pointed out that this was far too dangerous, as well as unnecessary (the fire wasn't spreading laterally at this point), and besides, they'd fall over on their own before long. His reply was that if the fallers refused to go in, he'd get the RCMP out there to order them in. I believe it was suggested that he go and screw himself. I know he left the site not long afterwards, surprisingly on his own accord. He did reappear on the fire a couple of times over the next few days, much like the occasional scavenging black bear and equally as unpopular. One of the latter pests eventually got into a loggers' crewcab and started going through one of the lunch pails. I only became aware of this when I happened to look up just in time to see the owner of the lunch racing off down the hillside to save the remains of it by chasing the bear off the landing. I'd never seen him move so fast—the guy wasn't exactly an energetic worker when it came to firefighting.

Not all of the logging crew resumed fighting the fire, as it turned out. Some of them carried on logging, presumably to ensure that merchantable wood was saved from the ravages of wildfire. On the day after the fire started, they managed to send eight truckloads of logs off to the mill. (Production before the fire was normally fifteen loads a day.) A few of the hollow cedar logs still had fire inside them when they were loaded onto the trucks, and they flared into flame as the trucks picked up speed driving down the road. The drivers had to return to the landing so the logs could be extinguished.

Another unfortunate event that took place during this fire occurred when I drank from a large juice can that had been sitting on the ground unattended; I was unaware that a wasp had crawled inside. It stung me on the tongue, which was downright painful. When I was talking to the first-aid man on the logging crew a bit later, I asked him what would happen if it had stung me in the throat and I stopped breathing. His reply was that I'd probably die; although he had an idea of how to do an emergency tracheotomy, he wasn't authorized to perform such a procedure. The only other first-aid case on that particular fire, as far as I can recall, was when this same first-aid man cut his hand with a small chainsaw and got me to bandage it.

It was on this fire that I got to set up a gravity water system for the first time. This involves running a hose line uphill to a water source (usually a small creek) to take advantage of the fact that every foot of elevation produces pressure of about half a pound per square inch. A gravity line provides water twenty-four hours a day and eliminates the need for a fire pump, which needs refuelling at regular intervals. The only downside on this particular fire was that the hose line had to be laid out straight up an avalanche slide area, which was covered in a thick tangle of slide alder and buckbrush. This vegetation is adapted to flatten down when snow slides over it, which makes fighting your way uphill through it rather like planing wood against the grain.

•

While we were finishing up on the last of the Smith Creek fire mop-up late one afternoon, I heard the sound of aircraft and looked up to see a large column of smoke rising above Double Eddy Creek a few miles away. Another fire had started on a logging operation, this time from a cable rubbing across a bone-dry rotten log, and once again the air tankers had been called in. I headed over that way to help out, but not before making a detour to the nearby logging camp to grab something to eat. By the time I arrived on scene the air tankers had finished working the fire and returned to base, as the light was fading. The fire was still very active, and due to the steep slopes and heavy timber, there was no sense in attacking it in the dark. A watchman was left on site to make sure none of the parked logging equipment burned up during the night.

The next day I was sent in as line locator to find a suitable route for an access road to one corner of the fire. It was hoped that it would be possible to lay it out so the grade would be suitable for hauling in water by tanker truck. Once that was done, I ended up waiting for a helicopter to come over and spot me on the ground for redirection in the heavy timber. One of the men from the logging crew was with me; it was the first time he'd been on a forest fire, and he was rather nervous about the strange noises going on inside the burn. I went to sleep by the edge of the fire since it was quiet at that point and I'd put in a lot of long days; besides, he was there to keep watch. I told him to shake me if the edge of the fire moved any closer to us. When I woke up later, he was still sitting in the same position, staring at the fire with his eyes wide open. The fire took a run shortly after that in the heat of the afternoon. We could see the black smoke through the trees and the occasional column of flame as the trees crowned. I think he found the accompanying roaring sound unsettling.

It was decided that an attack should be made at the head of the fire the following morning, and arrangements were made to send a crew in to set up a portable water tank above the fire that would be used to mix fire retardant, which would be pumped through a hose line down to where the fire was quietly simmering. Since we would be a fair way up the mountainside with no nearby water source, the only way the tank could be filled was by helicopter, using a monsoon bucket to dip water out of Kinbasket Lake and fly it up to us before too much leaked out. I was part of the advance party sent in to clear a spot for the helicopter to ferry in the rest of the crew and equipment. There were three of us: my supervisor, a faller and me. I don't think the faller had been in a helicopter before, as he seemed extremely nervous, which made me think I'd better keep an eye on him when we landed. There was very little room where the pilot set down, and in a situation like that the usual procedure is to go out the door, crouch down and not move. Trying to go in any uphill direction before the helicopter lifted off could be fatal, as you'd most likely be decapitated by the main rotor blades. Going in other directions would put you at risk of being de-limbed by the tail rotor.

The second the machine touched down, the faller wrenched open the door and was obviously about to get as far away from the machine as possible. I pounced on him and held him down until the helicopter had lifted off again. He repaid that favour by dropping a tree on me a few minutes later. It wasn't his fault: we only had a short time to clear a larger landing site before the helicopter returned, so we'd been worked feverishly. As the faller was cutting down trees with his chainsaw, my boss would push them in the right direction, and I would dart in to drag them away. Somehow the timing got mixed up, and a tree came down hard on the back of my neck. The impact stunned me, but I couldn't have been knocked out for very long, as they were lifting the tree off me when I came to. It wasn't a very big tree, and nothing seemed to be broken, so I picked myself up and carried on.

Eventually we also had a site cleared for the water tank, and once it was flown in, we set it up, along with the pump. A number of five-gallon pails of fire retardant were delivered, along with rolls of one-and-a-half-inch firehose. An access path was slashed down to the head of the fire to make it easier to lay out the hose (which was a good thing, as it turned out). Now all we had to do was wait for the water. It was a nice warm day and the view was spectacular, so we waited. And waited... Time passed by slowly, and we started getting bored with the view. When we radioed in for an update, we were told that the company was arguing with the Forest Service over who would be paying for the helicopter to make the expensive round trips to the lake. As the discussions back in town raged on, the afternoon got steadily hotter and there were signs of increased activity on the fire further down the mountainside. I made a few trips down the cut trail to see what was happening, and it was obvious that we needed to get water running through that hose line ASAP.

Finally the dispute was resolved and we were promised water forthwith, but unfortunately by then it was a bit too late, as the fire took a run uphill. I was at the end of the hose line when this happened, and I was almost bowled over as the faller who'd been working below me rushed past. He was a lot heavier than I was, and though I could run fast back then, there was no way I could keep up with him. We all gathered by the water tank to watch the show as the fire boiled up the mountainside past us, and discussed the illegitimate origins of the Forest Service personnel we held responsible for this situation.

We set up again the next day, further up the mountain, to try the whole operation over again. This time we did get the water we needed—it was delivered by helicopter as soon as the water tank was set up. The cost per gallon to fly water up to the fire must have been substantial, particularly after the monsoon bucket accidentally dropped from the helicopter a couple of days later, apparently due to an electrical malfunction of the cargo hook. It landed in heavy timber, which didn't do it much good.

During mop-up operations on this fire, we were getting water hauled up the fireguards in tanks secured to the back of rubber-tired log skidders. This configuration was referred to affectionately as a Ukrainian Water Bomber, in honour of the Ukrainian owner of the logging company that had been working the block where the fire started. An FMC tracked skidder had also been converted for hauling water in this way. It was slower than the rubber-tired skidders, but it and the operator were more stable. (Two of the other skidders were operated by brothers, who used to like racing each other on the landing when their boss wasn't around. Occasionally one would roll over, which wasn't a big deal, as the ROPS—rollover protective structure—canopies were designed to handle that kind of mishap. When their boss did hear about one of the rollovers, he merely asked the operator responsible if he'd checked the oil level before starting the machine up again once it was back on its wheels. When the operator admitted he hadn't, he was fired on the spot; however, he was hired back the next day, as he was a fast operator, which meant more logs dragged into the landing.) The FMC was used daily for crew transport, as it saved us having to climb up the mountainside first thing in the morning and back down again at the end of a long day. On one occasion there were eight of us riding on it: some on the tank, some on the roof and the rest hanging on somehow. It looked as if a Mexican bus had mated with a bulldozer.

Riding on the front of the rubber-tired skidders could be interesting, as you'd sit on the radiator and hang on to the canopy supports, taking care to keep your feet out of the way when the operator lifted the blade. One machine I was riding on, which had a thousand-gallon tank full of water strapped on the back, had defective brakes, as I discovered when we were going up a steep catguard. It was a noisy ride, as the exhaust stack was close to my ear, and I was hanging on tight to avoid being thrown off if one of the tires hit a stump or a large rock.

The trail got steeper and steeper, until suddenly the whole machine flipped up backwards and it felt like I was going up in an elevator. We would have rolled completely if the tank hadn't been attached. By attached, I mean that a couple of steel choker cables had been looped around it and connected to the winch hook. When the winch cable was tightened, the tank was pulled up tight against the fairlead. The only problem with that configuration is that each time the cable is tightened, it crushes the tank walls a little bit further, which reduces the amount of water it can hold.

We didn't roll, fortunately, and the operator got us back on all four wheels again. He started reversing down the trail, only to find the brakes weren't working properly, so we ended up travelling backwards faster than I would have liked.

•

We weren't the only ones to have a problem with the ground conditions. One of the sector bosses wanted a fireguard constructed straight down a steep slope. The soil in that location was what's commonly referred to as loonshit, as it has some similarity to that variety of avian excrement. There were several bulldozers working in that sector, but the operator of the first one that showed up refused to take on the job—there was no way he could back up once he started downhill. He rattled off elsewhere. The next operator who turned up similarly refused to risk his machine on a one-way trip, except not quite as politely. The third operator, who seemed to run his machine with a cigarette permanently attached to his lower lip, said he'd give it a go. I wasn't there at the time, but I heard the outcome on the portable radio I was carrying.

The bulldozer had nosed gingerly over the edge but didn't get very far before it began to slide, even with the brakes locked on, and it was obviously going to keep going all the way to the bottom. The operator and his cigarette bailed out rather than ride it down with the chance that the machine would turn sideways and roll. I happened to fly over that location the next day. You could see a long black streak where the blade had scraped all the way down, and at the end was what looked like a toy bulldozer from the height at which our helicopter was hovering. You could also see another bulldozer slowly pushing a rescue trail into it through the heavy timber. Amazingly, the runaway machine hadn't rolled; it had stayed right side up all its way down the six-hundred-foot trip. The scar it left was instantly named the Slick Slide (after its operator, Al Slick) and remained a good conversation piece for several years whenever company personnel flew over the site, until it grew over.

There were other hazards on the fire that were less obvious. On one memorable occasion one of the crew went off into the burn to take a dump while we were on a lunch break. This was perfectly normal procedure, and it wasn't until we heard a lot of swearing, followed by the unmistakable sound of a burned-out snag hitting the ground, that we realized something unusual had taken place. Our shouted inquiries concerning his safety were met by more unprintable language. When he rejoined us, he explained that the snag had fallen toward him while he was in a vulnerable position with his pants around his ankles. He didn't have time to pull them up, so he could only waddle like a duck as quickly as possible in what he thought was a safe direction. The snag twisted as it fell and once again aimed right for him, forcing him to change course. It definitely would have scared the shit out of him if it had all happened a few seconds earlier, he admitted.

●

We were finishing mop-up work on the Double Eddy Creek fire when we got a call that a third fire had just started a few miles away up Gold River. The cause was said to be friction from a falling tree, much the same way the Smith Creek fire had started a couple of weeks earlier.

We'd been fighting fire without a break since then, and most of the logging crews were getting fed up with it. I was perfectly happy to carry on as long as I could stay in the nearby camp and shovel down all the food I could eat, which was a lot in those days. I'd pack three separate lunches each day; it took a lot of energy to pack equipment up mountainsides. (Some time later I figured out how many calories I was consuming a day: it was close to ten thousand—which seems like a lot, but the US Forest Service suggests that its active firefighters consume at least seven thousand calories per day.)

I arrived at the new fire without any firefighting equipment. There was an FMC tracked skidder busy pushing fireguard around the fire, but a wind was blowing that was scattering embers into the forest and starting numerous spot fires. The FMC only had a small blade on the front that couldn't be set at an angle and was therefore not really designed for the task of building guard. Air tankers had been requested once again but hadn't yet arrived, so in the meantime three of us started hauling water from a nearby swamp in an attempt to extinguish the spot fires that were popping up all over the place. We only had two fire buckets, so the third man had to carry water in his hard hat. I discovered that we could double our efficiency by scooping up hot embers and dumping them in the swamp each time we returned for more water.

Everyone had to get clear of the fire area once the air tankers arrived, as it's not advisable to be underneath a retardant drop, particularly in heavy timber. I was in the process of checking that all the other firefighters who'd shown up were out of the way when the air tanker made its drop, much sooner than I'd expected. I'd barely managed to crawl under a log when the drop landed and covered everything in a slippery red coating.

This fire was a lot smaller than the previous two, but the company wasn't taking any chances by this point, so mop-up was long and thorough. We ended up with a fire boss who was a bit paranoid and wouldn't take into consideration the fact that it was starting to rain and the fire was virtually out. There's nothing more miserable than standing around holding a nozzle when you're soaked to the skin. By our rough calculations, the hose crews ended up pouring approximately 270,000 gallons of water onto a four-acre fire. It's a wonder we didn't wash the whole hillside away.

By the time things had settled down to where I was able to go back to town, I'd worked twenty-two days straight, many of them very long ones. I'd rushed out with only a toothbrush the day the first fire started, and I had literally worn the clothes I stood up in to rags.

•

The Square Triangle:

In forest fire control there is what's known as the Fire Triangle: Heat, Fuel and Oxygen (in the form of air). All three must be present for a fire to burn, and if any one of them is removed, the fire will go out.

It's suspected that there may actually be a fourth side that comes into effect on certain fires: Overtime. If allowed, fires can burn longer; if denied, it can lead to them being extinguished faster.

3
MAKING MORE SMOKE

By the end of the 1979 fire season, I'd decided that I wanted to stay in the forest fire control business permanently, even though I'd ended up having to drag a lot of wet and muddy firehose off the Smith and Double Eddy fires in miserable weather conditions. It turned out that all the hard work and long hours I'd put in on these fires had impressed Evans Forest Products, and they kept me on after the season ended, with the promise of making me a permanent employee of the Woodlands division, which was responsible for logging operations, in the near future. This meant I'd be able to participate in the slash-burning operations the company had planned for the fall. Going from putting fires out to lighting others deliberately sounded like an interesting switch.

Slash is the combustible material left scattered about on the ground after logging operations have been completed on a cutover area. It consists of a mixture of branches and foliage, treetops and logs that were deemed unsuitable due to decay or breakage. The distribution and concentration of this waste material will generally depend on the logging method used.

Conventional logging is done using bulldozers and rubber-tired skidders to drag felled timber to the landings along previously constructed skid trails. Landings are the levelled areas where branches and tops are removed, and the timber is then cut (bucked) into log lengths and loaded onto logging trucks. There are usually accumulations of slash resulting from these limbing and bucking operations that have been pushed out of the way to create debris piles.

High-lead logging is done with a cable yarding system that brings the timber from where it was felled to the landing where the yarder is set up, eliminating the need to build skid trails with bulldozers, which means less ground disturbance. Yarding with high-lead equipment makes it possible to log steeper slopes than would be feasible with conventional machinery.

The total slash fuel loading on a cutblock will depend on the species type and age and the health of the original timber stand. Logging decadent cedar-hemlock stands would result in the heaviest slash load due to the high percentage of trees that were unmerchantable due to rot. Fir-pine and spruce-balsam stands were generally healthier and there'd be less slash on the ground after logging, unless there'd been wastage due to disease or insects such as the mountain pine beetle or spruce bark beetle.

The traditional method of disposing of logging slash is to light it on fire, much the same way you'd dispose of a pile of branches in your backyard, but on a larger scale. This is called "slashburning," although the name seems to have been replaced by "prescribed burning." The latter term sounds more scientific, but essentially a slashburn is nothing more than a bonfire on a large scale—sometimes very large, if things go wrong. When an entire cutover area is burned at once, the operation is known as "broadcast burning," or simply broadcast. If only heavy accumulations here and there throughout the block are burned along with the landing piles, the process is called "spot burning."

Slashburning takes place at a time of the year when there's less chance that the fire will escape its boundaries. In the BC interior, broadcast burning is usually done in the fall, since by then the weather is getting progressively cooler and wetter, which helps keep the fires out of the surrounding timber. Eventually everything will get covered in snow, if worst comes to worst.

Before slash was burned it had to be isolated from the surrounding forest fuel to prevent what was politely referred to as "overachievement"—in other words, a burn getting away. This involved constructing fireguards with bulldozers, which are generically referred to as Cats, named after the Caterpillar tractor company. These machines were also used to pile slash, which usually involved reworking landing debris into piles so that the dirt mixed in with it was shaken out. This not only made them burn better, but it also lessened the chance that they'd smoulder under the dirt throughout the winter and into the following spring, to create a landing hangover fire. Cats would also be used to build more fireguard in the event of overachievement, at least until the ground became too steep for a machine to operate safely. On high-lead blocks it usually wasn't possible to construct fireguards for this reason.

I'd spent quite a lot of time in my first three years working fires watching Cats as they scraped their way around cutblocks, which did get rather boring—I was there mainly for safety purposes in case an accident occurred, such as the machine rolling over on a steep hillside. This never happened, fortunately, though I saw it come close a couple of times, and I would occasionally have to help extricate a Cat when it got stuck, by dragging the winch cable out to the closest stump that appeared sturdy enough to act as an anchor. Frequently they weren't, and the stump would be yanked out like a carrot, which was interesting to watch as long as it and the cable didn't fly in my direction.

I've always been interested in large machinery and enjoyed watching skilled Cat operators building fireguards and re-piling slash in landing piles in order to shake out the dirt. After I got to know one of the Cat operators quite well, we'd occasionally switch places when the machine had to be moved a mile or so down the road to the next work site. I'd drive his machine and he'd follow behind in my pickup, to give his back a rest. I'd have a great time working the steering clutches as I rattled and clanked along the road, hoping there'd be a tree blocking the way. Whenever that happened, I'd get the chance to push it off with the blade if it was a small one, or break it up if it was larger by running over it.

The sheer weight of these tracked machines could cause a minor problem, though. I was walking behind a Cat as it was making its way along some rocky ground on a hot August day when I noticed small puffs of smoke rising in its wake. When I investigated, I discovered that fragments of hot steel were coming off the tracks as they ground over the rock and starting fires in the bone-dry ground fuel. I stamped them out easily but decided I should have a backpack water tank with a hand pump (known as a "piss can") on site to deal with any other fires that might accidentally spring up. Accordingly, I radioed the office to inquire if there was anyone coming out that way who might be able to drop one off. The result was unexpected: they sent out a twenty-five-hundred-gallon tanker truck, along with two additional men, three piss cans, two shovels and two Pulaskis. If they'd just sent me a case of beer instead, I'd have been set up to take care of extinguishing further fires on my own.

It was possible to make the job of Cat supervision more entertaining, as two of us did when we drove out to check on an operator who was windrowing slash (piling it in long rows) up the Blaeberry Valley, northwest of Golden. The cutblock he was working in was on the other side of the river and there was no bridge anywhere near, so we waded across, which wasn't a great hardship, as it was a hot afternoon in the middle of summer. Once we reached the other side, we spotted a dead spruce tree at the top of the riverbank and decided that we'd light it on fire as a surprise for the Cat operator. We were below the level of the ground he was working on, so all he saw was a tree seemingly bursting into flames spontaneously. Maybe he was expecting to see someone with a white beard suddenly appear with a couple of stone tablets. We were enjoying his reaction until I suddenly realized that the afternoon air-patrol plane would be due over the valley shortly and could hardly miss the column of black smoke we'd generated. There was no possibility that our little joke would cause a forest fire, but the Forest Service wouldn't have been impressed by the fact that the tree had been deliberately torched. The spruce was still burning merrily as I waded back across the river to fetch a chainsaw from the truck. We cut the tree down and then went over to get the Cat to drag it over to a gravel area and bury the incriminating evidence, reasoning that any smoke still hanging around when the plane arrived would be attributed to exhaust from the machine.

It's slow and costly to have bulldozers walk long distances under their own steam, so whenever they have to be moved to a new work site they're generally transported by lowbed. Quite often I'd have to meet a lowbed at a designated spot and then pilot it in to where the Cat was to be dropped off. Logging roads in the mountains are usually narrow, with few places where a lowbed can turn around, and there are often sections of road or bridges that won't bear the weight of a lowbed with a thirty-ton bulldozer on board. Sometimes a Cat would be delivered very early in the morning, which would mean I'd have to leave town even earlier. This could be stressful after a long night spent in the bar or at a party, and on at least one occasion I drove straight from a party late at night and parked my truck so it blocked the logging road at the rendezvous point. I was awakened after an indecently short interval by the rumbling of a diesel engine and looked up to see the lowbed idling inches from my bumper, with the driver grinning from ear to ear. At least he'd been considerate enough to not use his air horn.

There were some hazards involved in supervising fireguard construction. In addition to avoiding steel cables flying through the air, there was always the danger of being in front of a Cat as it was working its way through standing timber. In that situation the operator is unlikely to see you up ahead as he's pushing over snags or trees, and when they are knocked over by a Cat blade, they fall much faster than if they'd been cut with a saw. I've seen a couple of near misses when this has happened—on one occasion the tree came so close to one firefighter that a branch knocked the radio off his belt.

One of my near misses happened when I was scouting fireguard location just ahead of a Cat at the back end of Bachelor Creek. While I was standing out on the end of a log, looking at the swampy ground below and scratching my head, wondering if the heavy machine would sink in the bog, I felt the log start to move. I instantly realized that the Cat had caught up with me faster than I'd expected and was pushing against the other end of the log, and that very shortly I'd be flipped off into space. I looked down to pick a reasonable landing spot and then jumped off into the dubious safety of the swamp just in time.

The standard way to approach a bulldozer working in the timber is to come up from behind and throw a chunk of wood against the back of the protective cab. This lets the operator know you want to talk to him. It's important not to throw the chunk from the side, because if you hit the operator, he'll definitely want to have words with you.

•

My first experience with industrial slashburning began in the fall of 1979 when I was sent to supervise burning operations on the company's timber holdings near Revelstoke. The slash fuel was too damp to get any broadcast burns going, so we concentrated on lighting heavy accumulations within the cutblocks and piles at the landings. There was a labour shortage in town at the time, as it seemed everyone was working on the Revelstoke Dam construction project. Fortunately there was a contract burning crew available to help, and I was able to hire a few additional bodies from the town's female population. The latter group included a cocktail waitress who'd never worn workboots before and found them really tough on her feet, as well as a sixteen-year-old model with long blonde hair that kept getting singed during lighting operations. They both worked quite well despite these handicaps, although I don't think either of them took up slashburning as a career.

The contract crew was from Salix Resources, which was actually IFFS, my former employer, under a new name. Due to financial upheavals, the company had been going through names faster than I'd been going through shirts, and the details were equally grubby. I was glad to see some former acquaintances, though, and burning operations proceeded amicably and without major incident, apart from the Salix crewcab nearly burning up. It had become stuck in the mud beside a burning slash pile and proved so difficult to extricate that by time I eventually managed to tow it out, the two fuel drums in the back were beginning to smoke from the intense heat.

This was the first time I'd dealt with the Forest Service regarding burning permits, and I found the Revelstoke personnel really helpful and cooperative. They weren't sticklers about burning permits, and on one occasion I had a forest officer writing out a permit on the hood of my truck for some piles that I'd already lit, as we were being showered with sparks and burning embers. It was also the first time I met Puff the Magic Dragon (a piss can full of slashburning fuel that had been adapted for use as a flamethrower). I liked it so much I made up my own as soon as I got back to Golden.

During this time, I stayed at McGregor's Motor Inn, where there were tile floors in the lobby that caused problems, as I'd used wood screws to reattach the soles of my disintegrating work boots. Every time I entered the place, the protruding screw heads would skid on the tile surface, leaving the occasional gouge mark. My jacket smelled strongly of diesel fuel, which made me even less popular, and the desk clerk would ask me pointedly every morning how soon I'd be leaving their establishment for good.

Rain put an end to the Revelstoke burning project, so I rushed back to Golden, hoping that it wasn't raining there so that I'd be able to take part in the light-up operations at that end. Unfortunately the rain got there at the same time I did, which put an end to any further broadcast burning in the 1979 season.

•

The following year was much better. I'd just returned from spending nearly a month supervising road construction and camp set-up at Tsar Creek and was sitting in the pub having a beer when one of the Evans Forest Products staff joined me at the table. Apparently there'd been a reorganization of the Woodlands division during my absence, and I was now the new fire protection officer. I thought he was joking at first, but it turned out to be true. I suddenly found that I'd have to conduct all the upcoming slash-burning operations, and as my predecessor had left things somewhat up in the air, I'd first of all have to prepare the necessary burning plans. My only experience to date had been the spot burning I'd supervised in Revelstoke the previous year. I'd never been on a broadcast burn before, so I had never witnessed how they were actually carried out. I did, however, have a copy of the BC Forest Service's *A Guide to Broadcast Burning of Logging Slash*, a twenty-page pamphlet that had a helpful sample burn plan diagram in the back. Unfortunately this plan was for a near-flat block, and there weren't too many of them on the list of blocks that I was expected to burn in a few weeks' time. I thumbed through the pages several times nevertheless, until I felt a bit more comfortable with the theory.

The first block I chose to burn was a flat one in the Beaverfoot Valley. Once I'd decided that the slash was dry enough to burn and had picked an auspicious day with a forecast of cooperative weather (i.e., no howling winds forecast for at least twenty-four hours), I gathered all the tools together. Tanker truck, standby Cat, helicopter, crew (only a couple of them were tools), driptorches, fuel and several copies of the burning plan I'd drawn up—it all seemed to be there. I'd forgotten my matches, but fortunately all the crew seemed well supplied. As I leaned on the hood of my truck, contemplating the chances of accidentally burning the surrounding forest if I screwed everything up, the crew helpfully suggested that now would be the ideal time to start lighting. I was new to the situation and hadn't yet figured out that perhaps some of their advice should be considered carefully. They'd all had previous experience in broadcast burning and were obviously enjoying the spectacle of a new burn boss on his first day. I can't hold that against them; in their place I'd have probably done the same. Eventually I decided the conditions were right and we went ahead with the light-up. Surprisingly, everything went off well and there were no escapes.

After I'd lit up a few more blocks, I began to get used to broadcast burning and the peculiarities of the burning crew. It didn't take long to figure out that they weren't too upset if a burn happened to spread out of bounds, as that meant more overtime. I never caught anyone deliberately scattering fire on the wrong side of the fireguard, but I am certain that it happened more than once.

I'd draw up a burning plan for each block, with little red arrows to show the planned ignition pattern and big black arrows showing the escape routes in case things got out of hand. A copy would be sent off to the Forest Service for their approval, but I never had anyone there dispute a plan or ask questions. I was tempted to draw one up with the escape arrow pointing directly into the burn, to see if anyone noticed, but decided that would be a little unprofessional. I'd make copies of each plan for the burning crew and hand them out on the site so they'd be aware of the escape routes, but they either threw them away instantly or saved them to use as toilet paper later on. Quite often I'd alter the ignition pattern due to changes in weather or fuel conditions once light-up started, which rendered the plan inaccurate anyway.

There was generally a lot of anticipation when we were waiting for the moment to begin lighting. The crew would be sitting around on the fireguard with their driptorches at the ready, and the helicopter pilot would be waiting for the word to crank up his machine. Wind was always the big unknown, particularly in the mountain valleys. We'd try to take advantage of the downslope winds that would start in the late afternoon and evening, but sometimes they weren't reliable. To test which way the air was moving, we'd light a small fire on bare dirt and watch which way the smoke was drifting. In towards the block was good, but back the other way into the adjacent forest was not good at all. You could get the same idea by picking up a handful of dust and letting it trickle out through your fingers, but test fires were better, as you could toast sandwiches while you were waiting.

Broadcast burn light-up was done with hand and/or helicopter ignition. Hand ignition was carried out by walking through the slash while holding a driptorch, so that you left a trail of burning fuel in your wake. The helicopter had a much larger version, with forty-five gallons of fuel slung underneath, and would light from a higher altitude. There were two main ignition methods: strip firing and convection burning. The first method was suited to steep ground, and it involved lighting a strip along the top of the slash block close to the fireguard then letting it burn slowly downhill. This in effect was using fire to widen the fireguard. A second strip would be lit parallel to the first a bit further downhill, and the two lines of fire would be allowed to join up. Once it was considered that there was enough of a burned-out buffer at the top to make it safe, the rest of the block would be ignited in strips, either by hand if it was a small area, or by helicopter.

Convection burning involved lighting the centre of the block first and then working out concentrically toward the perimeter. The idea was to take advantage of the indraft generated by the fire, using it to draw each ignition line into the central fire. This method was ideal on flat ground or where there was a ridge inside the block being burned. It could also be used on sloping ground, depending on slope angle, fuel loading and other factors.

Sometimes we had to wait until late into the evening for the wind conditions to be just right. This caused a problem if we were using a helicopter, as it would have to be back at its base by what was known as Legal Grounding Time. We were often working a long way from town, so we sometimes would end up having to let the helicopter go and finish lighting by hand. One of our burns at the back end of Bush River finished so late that the pilot ended up parking his machine in the woods and staying the night with us at our makeshift camp because it was too dark for him to fly home.

Hand lighting in the dark was quite often entertaining, as you had to keep track of where the other members of the crew were so you didn't trap them with the line of fire you were leaving in your wake. Walking through slash could be tough in the daylight if it was particularly heavy, but navigating it at night took a bit of getting used to. One night I'd walked along a log with my driptorch as I was looking back at the burn's progress and suddenly found myself at the end of the log looking down into space. The ground had dropped off and it was too far to jump down, so I had to go back the way I came. By this time the fire I'd dripped had taken hold in the slash, so as I walked back along the log there were flames all around me to make balancing on it more interesting.

Once light-up of a block was complete, the next stage of the operation was control. If the burn had gone well and no fire had jumped across the perimeter fireguards, this would be the time to sit back and have a beer while we monitored the situation. If there was an escape, we'd take suppression action as long as it was safe to do so. Hoses would be strung out to bring water to the trouble spots, and possibly the standby Cat would be set to work building guard to cut off the fire's spread. If it wasn't possible to take immediate action on the escape due to safety concerns, the best thing to do would be to open another beer and plan strategy for the following day. There would be an inverse relationship between the number of empties lying around the site and the success of the burn.

Control problems would put a stop to any further block light-up, and we tried to avoid getting into this situation. Whether there was an escape or not, some burns would require a certain amount of mop-up once everything was under control. This was the least popular phase, as it could be slow and dirty work dragging hoses around to extinguish any hot spots within the burn that might cause problems later. Accumulations of fuel just inside the catguards that hadn't been consumed by the fire were always a concern and would be lit with driptorches to burn them completely in order to avoid them flaring up later and sending wind-borne hot embers across the guard. This is why Cats constructing fireguards always tried to set their blades so as to push debris outwards, to avoid leaving slash mixed with dirt on the inside of the perimeter. It wasn't always possible to do this along the top edge of a block on steep ground.

Even after a thorough soaking with water from hoses (or a convenient downpour), there would still be hot spots in the duff ground fuel that weren't putting out enough smoke to be obvious. Nowadays infrared scanners can be used to detect these problems, but the traditional method of locating hot spots is what's known as "cold trailing." This involves testing every inch of the ground with your hand, the idea being that if you burn your fingers, you didn't do a thorough job of mop-up. I've caught individuals wearing gloves while cold trailing, presumably to protect their delicate skin, and I have suggested that perhaps they might be more suited to hairdressing than firefighting.

Our broadcast burns inevitably generated smoke, and the larger ones (up to 250 acres) could create smoke columns that were thousands of feet high. We managed to smoke out the Trans-Canada Highway for a couple of days when we were burning blocks at lower Quartz Creek and Beavermouth. We did the same to Revelstoke when we were burning just upriver above the dam site, only this time it was for nearly a week due to weather conditions. Strangely, no one in Revelstoke complained, probably because it was a sawmill town and the residents knew that slashburning was part of the logging process. It was so bad some mornings that you could almost cut the air with a knife.

When we received complaints from Calgary about smoke from our broadcast burns in the Beaverfoot that had travelled east and was spoiling their air quality, we simply blamed it on burning being done by another timber company operating in the Revelstoke area. No doubt when contacted, they in turn blamed it on yet another company further to the west.

•

There were never enough windows of opportunity to burn all the blocks we had scheduled, due to the weather being uncooperative. Slash would either be too dry or too wet at the start of the fall broadcast season, which was traditionally right after the September long weekend. Too dry meant the burn would most likely get away, and too wet meant we'd have a hard time getting anything to burn at all. When conditions did get just right, we'd burn one block after another until either the weather broke or we had escape problems that forced us to stop. After I'd been burning for two seasons, I decided to try broadcast burning in the spring as a way of getting more acreage prepared for planting seedlings. Spring burning can work well for blocks situated on south-facing slopes, as the slash can become dry enough to burn while there's still snow on the ground in the forest above. It poses additional risks, however, as unlike fall burning, the weather can get hotter and drier as the weeks go by and spring moves into summer. If you haven't put out every vestige of fire on the burn, it will flare up later and cause real problems.

I was fortunate with the first spring burning I carried out in May 1982 at the top end of Copper Creek, for there was very little overachievement and nothing came to life at an embarrassing moment later on that summer. There had been no catguards on any of the steep high-lead blocks, but there was enough snow on the ground under the trees surrounding the slash fuel to keep the fire confined. This was due to the fact that the solar radiation that had dried out the slash nicely on the steep south-facing blocks hadn't penetrated into the adjacent timber. I hoped these ideal conditions would occur every spring (they didn't). The burn was very hot; in fact the helicopter pilot reported that radiation from one of the blocks he'd been flying past after lighting was so intense, he'd had to fly with one hand shielding his face, and he was concerned about his paintwork.

The burned blocks were patrolled every day, even after it appeared that the last of the hot spots were completely out. Spring broadcast burns consume less of the duff than fall burns, since only the topmost layer has had a chance to dry out, and the remainder underneath is still saturated from over-winter moisture. The weather can get hot and dry quickly in the mountains as spring progresses, and it can cause the lower duff layers to dry out rapidly. The entire block may appear black and dead with not a trace of smoke, even after many daily patrols, but there may still be a glowing ember or two hiding in the duff. I'd been walking through one of the burned Copper Creek blocks one afternoon and hadn't seen any sign of smoke, but when I stopped halfway across and looked back, I saw a patch of burning duff approximately ten feet square right in line with the route I'd taken. The wind and low humidity had suddenly brought an ember back to life.

•

The following year I planned to do more spring burning and picked a few blocks in the back of the Glenogle Valley as being suitable candidates. The only problem was that in order to be able to burn the blocks when conditions were right, we had to open up the road to the back of the valley. This involved a week of plowing with a D8 Cat, which was easy at the start of the road where the snow was only a foot deep, but harder further in where the snow was over seven feet deep. The road crossed a number of avalanche zones, which posed problems: not only did we have to be careful working under the slide path, but the Cat operator had to make sure his machine was tilted at an inward angle toward the hillside. If he was angled the other way, there was a chance the whole machine would slide over the edge as the smooth steel tracks slipped sideways on the packed snow. We made it to the blocks without mishap, but as it turned out the weather didn't cooperate—by the time the slash was dry enough to burn, the adjacent forest was too dry to risk lighting up, so nothing was burned except the money spent on plowing.

The broadcast burning program was to a great extent driven by the requirements of the silviculture department. Seedlings had been ordered, and I had to prepare the ground for planting them by burning up the slash. It's not always possible to find alternate places to plant seedlings if the site can't be burned due to uncooperative weather, as they can only be moved within a limited range of aspect and elevation. A few blocks shouldn't have been burned because the risk of an escape was too great. I lit them up with misgivings, after warning the silviculture department that there was a good likelihood of the fire taking off. The burns were requested nevertheless, as seedlings were on their way, so the blocks were burned—along with part of the mountainside above, unfortunately.

Some blocks were at high elevation, and they had to be burned before the snow covered them up, which meant starting on them in the last part of August. A good snowfall would help with mop-up operations, but if there was too much of the stuff, there was the chance that we couldn't get back in to pull all our firehose off the site before winter set in.

In order to deal with the backlog of blocks needing burning and to make room for the increasingly ambitious replanting program, it was necessary to extend the burning window even further, and accordingly the decision was made in 1986 to try a summer burn in mid-August. My first choice of location was upper Bachelor Creek;

winter often came early in that part of the country. I had to rule it out, however, as the Forest Service was busy fighting a wildfire in the same valley that August, and I didn't want to add to the smoke they were already dealing with. I also thought there was something slightly wrong about lighting fires nearby at the same time as the government was spending the taxpayers' hard-earned money on putting one out. I picked Marl Creek as a suitable site instead; there were several fairly level blocks with good road access and water sources nearby. But things didn't work out exactly as planned.

•

"Kindle not a fire that you cannot put out."
—Old proverb

4

FIRE MISBEHAVIOUR

Escapes took place occasionally during broadcast burning operations, and depending on the time of year the burn was conducted, they could prove to be quite unpleasant. I did my best to avoid escapes, so I would go to a lot of trouble monitoring fuel and weather conditions in an attempt to get a burn that would achieve the desired silviculture objective (plantable ground for seedlings) yet stay within the prescribed boundaries. I'd set up weather stations out in the woods as soon as the snow melted enough to give access to the sites. The data collected was used to determine optimum burning conditions as well as for keeping track of fire hazard for the company's logging operations during the summer.

When we were on a block getting ready to burn it, I'd use the radio telephone to get spot forecasts from the Castlegar weather office. The further we were up the mountain valleys, the more unreliable these forecasts were, but it wasn't the fault of the weather technicians.

Mountains can make their own weather, and we sometimes had to contend with strong downdrafts suddenly appearing out of nowhere, particularly if we were close to an icefield. One afternoon up Bachelor Creek I was listening to the man at the weather office explain that we should be experiencing calm conditions where we were, as a howling wind was meanwhile blowing all the miscellaneous papers that had accumulated on the floor of my truck out the open doors. Normally I wouldn't have minded—it saved me shovelling out the vehicle—but my burning plans were blowing away with the rest of the trash. Needless to say, that particular burn was postponed for the day.

On other occasions the wind would show up unexpectedly after we'd started lighting and bend the column of smoke and flame toward the timber, which usually led to the burn spreading beyond the prescribed boundary. Anything less than a couple of acres was considered "fringe damage"; over that was deemed to be an escape. I tried not to have our overachievement labelled an escape whenever possible, as this avoided a lot of extra paperwork back and forth to the Forest Service. Besides, we'd have had to traverse the escape area, and sometimes that would have meant crawling about on very steep terrain. Generally, the Forest Service agreed with me, as their staff didn't relish the prospect of climbing up the mountainside to check things out either.

Now and then I'd cancel a burn if I had severe misgivings about the weather. It was a difficult decision if there was a lot of equipment gathered on the site ready to go, since it was all costing money, as was the manpower standing around. Sometimes, though, it was better to write off the costs and give up for the day as opposed to lighting up and having a really expensive next few days if the weather did turn out nasty. I didn't like it at all, however, if a burn was cancelled due to human stupidity, as in the following instance.

As we were getting ready to burn a block in the Beaverfoot Valley, I gathered from chatter on the company radio channel that there were a couple of goat hunters missing in the mountains west of Golden. The helicopter I'd be using for light-up later in the day was going to be involved in the search, so I radioed a request that a backup machine be available for my use that afternoon if needed, which was agreed to. It turned out that I never did get my helicopter that day, as it was still tied up rescuing the two idiot trophy hunters from a precarious ledge that they'd become stranded on in their attempt to climb down to retrieve a mountain goat they'd shot. The backup machine was never ordered, so I was forced to cancel the burn, and my comments over the radio about goat hunters were apparently quite enjoyed by the RCMP officer who was riding in the rescue helicopter. The stupidest part of the whole business was that the dead goat was left on the mountainside, as it's apparently illegal to pick up game with a helicopter. I flew over the animal a day or two later, and the sight of it only deepened my dislike of trophy hunting.

Strange as it may seem, the company generally wasn't that upset when a burn wandered off into the surrounding forest, as long as it wasn't a major disaster, because they could go in afterwards and harvest any scorched trees that were of merchantable size. This led to an odd situation the day after we'd burned a block at Beavermouth near the eastern end of Rogers Pass. The company logging supervisor was quite annoyed because the burn hadn't escaped as he'd expected and hoped, since there was a prime fir stand above it. The burn had been a difficult one due to heavy fuel loading on the block and fire-induced winds that at one time were strong enough to lift fine gravel off the road. I thought we'd done a good job by holding the burn within the boundaries, but apparently we'd achieved the wrong result.

Sometimes I was needlessly concerned about negative feedback. When our slashburn on a block immediately adjacent to Kootenay National Park escaped across the boundary a short distance, I called their headquarters rather apprehensively to explain what had happened. Surprisingly they weren't too upset, and they gave us permission to send our bulldozer into the park to construct a fireguard around the escape, as long as we did the minimum amount of damage. They sent a park warden out to monitor our activities, but he was so unconcerned that he just slept in his truck and never once got out to talk to us.

At times our burns were actually quite popular with the public. When we were planning a burn at Palliser, just east of Golden and opposite the Trans-Canada Highway, we were told in no uncertain terms that we were not to let it get away, as it was the year of Expo 86. The Forest Service was deeply concerned by the prospect of all the people driving by on the way to or from this event seeing Beautiful BC being set on fire deliberately. They were so concerned in fact that they stationed one of their staff at the side of the highway to explain to anyone stopping that what they were witnessing was not actually a forest fire but a controlled burn. We needn't have worried, as it turned out: many of the tourists who stopped to take photographs while there was a bit of fringe damage taking place exclaimed that they were the best pictures they'd taken on their entire trip.

Wind wasn't always to blame for a slashburn getting away; sometimes the problem was the nature of the block and its surroundings. There were some blocks that really shouldn't have been burned, due to steep slopes and spruce bark beetle–killed trees above them just waiting to catch fire. Even if the fireguards held, burning debris lifted up in the convection column would be dropped into the standing timber, and once a few trees ignited, the fire would carry on up the mountainside.

One block we burned in Glenogle Valley was rather difficult to light up, as the ground was too steep to construct a fireguard along the top perimeter. In fact, it was so steep that the light-up crew had to hang onto things with one hand as they picked their way along with their torches, lighting slowly and carefully. I never expected we'd be able to keep the burn from escaping, and I had passed my concerns on to the silviculture department, but they gave the go-ahead because they really needed ground prepared for planting seedlings. It really wasn't much of a surprise when the fire went crowning up the mountain. At least this burn took place in October, so it wasn't too long before snow arrived to help us put out the escape.

Escapes resulting from spring broadcast burns were a lot more trouble to extinguish; usually the weather would disconcertingly get hotter and drier when we were hoping for a nice heavy rain shower. At least, I'd be hoping for a downpour; the burning crew would be perfectly happy if the fire burned on for the next few months so they'd get the necessary number of weeks in for their usual off-season UIC claim.

•

The first spring burn I'd undertaken was on the high-elevation Copper Creek blocks mentioned earlier. That 1982 operation had been carried out in spruce-balsam slash and the results had been satisfactory, with fairly easy mop-up. That wasn't the case when we did some spring burning up Bush River the following year. The main problem was that I'd failed to take into account the fact that we were in a different fuel type: cedar-hemlock slash, with many of the waste logs scattered throughout the slash being hollow. Slashburns don't consume every bit of fuel on the ground, or at least they're not supposed to, as that would end up cooking the soil and degrading the site. Logs end up charred and partially consumed, and any residual fire on them can be put out with a hose or a few good downpours. But hollow logs will keep on burning inside even if it rains for several days, and we ended up having to cut them into sections and split them open so that we could hose down the insides. I made a mental note to avoid burning decadent cedar-hemlock blocks in the spring from then on. The other problem was the weather: we kept getting hot, dry winds and very low humidity, which didn't help our firefighting activities in the least.

It wasn't just the logs on the ground that caused us problems at this burning operation in the Bush River valley—most of the standing cedar trees were hollow as well. Falling these when fire had got inside them was a nightmare, and the qualified faller we had assigned to us refused to take on the job, saying it was too dangerous. I had to get the trees taken down to make the area we were mopping up safe to work in, so I persuaded him to let me use his chainsaw. He was understandably reluctant, figuring it would get destroyed when one of the trees inevitably fell on top of me. I assured him that the company would buy him a new chainsaw if the worst happened, and I started cutting. I'd never been trained as a faller, apart from the instruction I received during the training I did when I was hired on the IFFS crew six years earlier (a total of about thirty minutes of direct instruction). The hollow cedars were dangerous and unpredictable, as there was very little shell wood holding them up, and if you weren't careful, the whole tree would slide down onto the saw if you cut too far through the shell. To make it even more interesting, they were burning inside all the way to the top, like a chimney fire, and once I cut through the shell, flame would come out and make the front handgrip too hot to hold. I ended up using my left foot to push on the handgrip. Eventually I got them all down without having an accident, although the heat from the inside of the trees ruined the bar of the saw. It was after that horror show that I started using explosives to take down burning snags and hollow cedars, fully aware that if I continued cutting in this manner it would only be a matter of time before one of them got me. We finally extinguished the last of the hollow cedars without any loss of life, but there were a few close calls.

It was another six years before we conducted the next spring burn operation up Bush River. This time we were higher up the valley in spruce-balsam forest, so things should have gone better. It turned out they were in fact much worse, and I began to think the valley had some kind of jinx. The burning seemed to go well at first, with very little in the way of fringe damage, which was good, considering the ground was so steep it wasn't possible to construct fireguards, and we were relying on the residual snow in the timber above the block to hold the fire.

My old nemesis, the Bush River wind, struck once again, this time with a howling gale that came out of the mountains at the head of the valley and stirred up the fires smouldering within the blocks, to the point where the whole lot was on fire once again. The wind was so strong that as I drove at high speed from camp toward the growing smoke column, I witnessed a sizeable spruce tree beside the road being snapped off. This turned out to be the largest escape I ever had during the years I was slash-burning, and the control operations went on for some time, as the fire had gone right up the mountain.

One of the biggest problems was that the fire wasn't easily accessible, as the bridge across the river had been removed. The blocks we'd just burned had been logged in winter, and the bridges were taken out so that they wouldn't get washed away when the river rose in the spring. At first we found the water shallow enough that we could cross in pickups, or wade, though the water was very cold—we weren't far from the icefields. As the days passed, the river rose slowly and steadily thanks to the warm spring weather. To make life more interesting, the water would rise even more during the afternoon when the volume of meltwater approached its daily peak. I found this to be a problem the day I waded into the river only to be swept off my feet and washed away downstream. By the time I made it to the other side, I was quite a long way downriver from my starting point.

When the water rose even higher still, it was no longer possible to get across on foot or in a pickup truck, so we travelled over in style atop a rubber-tired skidder. A couple of the Forest Service staff came out one day to take a close look at our mop-up operations and seemed horrified at the method of transport we were using. They'd brought along an aluminium dinghy, which they planned to use to get across the river. I warned them that the current was too strong for paddling across, and that they were more than welcome to ride over with us on the skidder. They refused, saying that this would be a direct violation of Forest Service safety protocols, and proceeded to go about launching their boat.

The crew and I watched with interest, particularly when the current yanked the boat downstream so violently that the man holding the rope attached to it was nearly pulled into the river. After some consultation the pair walked over to us and asked furtively if they could ride across with us on the skidder after all. I pointed out that this would surely be a contravention of their safety guidelines, but as I recall, their response was, "Fuck the regulations!" We loaded them up on the machine with the rest of the crap we were taking over, and then the skidder charged off into the river, with all its passengers hanging on tight.

Soon after that we managed to drag a log across a narrow part of the river with the skidder in order to serve as a precarious footbridge for gaining access to one of the blocks. I crossed over this barefoot early one morning, with my boots hung around my neck and a sack containing sticks of dynamite slung over my shoulder. When we got to the other side, we were met by a US firefighting crew that had been sent up to help as part of the Interagency Mutual Aid Agreement. They had been flown across the river by helicopter—a crazy thing to do in my opinion, as the valley was choked with smoke from an overnight inversion, which made for lousy flying visibility. I told them so and asked them why they hadn't just walked across the log like we did twice a day. They replied that OSHA regulations wouldn't allow them to do anything that hazardous. The worst that could happen to us would be to fall in and get wet. The worst that could happen to them on their morning flight would be to crash and burn.

They asked one of my crew who was in charge, and he pointed to me, standing there barefoot in ragged, filthy clothes that hadn't been washed in weeks. The next question was, "What's in the sack?" Dynamite, they were told, and this got them really excited—as they explained, they weren't allowed to use explosives, and could they come with us and watch it being set off? I had to disappoint them.

•

We were short of firefighters one day during mop-up operations on the part of the escape we were working on. The Forest Service was doing head counts each day to ensure we had the correct number of men on site, as per the written instructions we'd been given, so we urgently needed a few more (live) bodies. On the way out to the site I spotted two characters fishing in a creek beside the logging road, and stopped to see if they'd like to work for us for a day or two. They seemed happy to come along and make a bit of money, and they followed behind me in the dust, which was good, since I had legal authority to apprehend them for firefighting duty if they'd refused to come along.

I should have left them where they were, as it turned out, as one of them proved to be a completely useless bastard who had serious drug problems and could literally not be trusted with anything sharp. After some hair-raising incidents occurred while he was working with the mop-up crew, I finally sent him back to the trucks, with strict orders to stay there. If a Forest Service vehicle showed up, he was to pretend that he was busy filing something, just as long as it wasn't an axe, as he'd probably lose a finger or two. We booted him and his partner off the site as soon as we got two suitable replacements.

Once the surface fires were out, we had to deal with the landing fires that were still burning underground—if they weren't extinguished, they could pop up later in the summer and start the conflagration all over again. We ended up digging down a long way with a Cat so that we could hose everything down, extracting whole logs that had been turned into perfect charcoal in the process. One of the crew happened to have his ten-year-old son with him. This kid was actually the hardest worker on the site, but I couldn't put him on the payroll, as he was somewhat underage. His father was therefore given extra hours that equalled what the boy would have earned. The company would have had a fit if they'd known about it, but we got the job done, and that was always the bottom line as far as I was concerned.

One Friday afternoon, after we'd been working on the Bush River mop-up for quite some time, it became obvious that the crew was getting a bit fed up. We'd been staying in the logging camp a few miles downstream all this time, and they were getting antsy, particularly since the supply of alcohol they'd brought out with them had completely run out. They wanted to drive all the way back to town for the night and promised that they'd return early the following morning.

I knew exactly what would happen: they'd close the bars down and be too pissed to remember where they were supposed to come back to, let alone when. The company was under a legal obligation to have a minimum number of men on the mop-up operation, and the Forest Service were watching us very closely to ensure compliance, I explained to the crew, adding that under the circumstances, I couldn't let them leave camp. They all swore that they'd behave themselves and go to bed early, but I didn't believe a word of it. Eventually I told them that I was going to remove the keys from the vehicles so that they couldn't drive to town, and their response was that they'd either take the keys off me by force or hot-wire the trucks. I told them that if either of those events took place, I'd be forced to radio ahead and have them arrested by the RCMP for leaving a fire without permission, and they'd be sent straight back under escort. The crew stayed under protest, but I kept a wary eye on them that weekend. Eventually all the fires were completely extinguished and most of the people involved were reasonably happy.

•

Our 1986 summer burning operations at Marl Creek, on the other hand, made quite a few people distinctly unhappy. This was the summer burning operation we'd planned for mid-August, and it seemed everything was progressing well at first, and all the blocks were lit up without incident. During the mop-up phase, however, high winds that hadn't been predicted in the weather forecasts came howling across the blocks that were still smouldering, causing them to flare up, and fire jumped the catguards into adjacent standing timber. The ground wasn't steep enough for it to make a major run, but there was so much smoke that we were unable to attack the head of the fire. All we could do was watch as flames spread into a stand of timber that had been selectively logged some time earlier and hope the wind died down. There were young plantations and a sawmill downwind, so the air tankers were called in, but for some reason the bird-dog officer who accompanied them completely ignored the men and equipment gathered on site, and he radioed back a report that the head of the fire wasn't being adequately attacked. How he expected us to work in heavy smoke I'll never know, as unlike structural firefighters, we didn't have any breathing apparatus and therefore wouldn't have survived very long working in those conditions. (The only smoke-filled environments slashburners seemed to thrive in were the local bars, and the inside of the crewcabs when the windows were rolled up.)

The atmosphere got even more unpleasant when the Forest Service announced they'd be taking over all fire control operations, a decision I disagreed with strenuously. It seemed they'd believed the bird-dog officer's report, even though by this point the wind had eased off and the crews were able to resume attacking the fire. When the senior company representative who'd arrived on site agreed to relinquish control, even though I warned him it was unnecessary and would set a bad precedent, I got fed up to the point where I resigned on the spot. Word of this spread quickly, and apparently not long after I left the site, some of the crew quit as well. I'd like to think this was out of loyalty, but I suspect that perhaps they felt a Forest Service fire boss might be somewhat less tolerant of any peculiarities they might have.

Later that evening my supervisor tracked me down in the pub as I was planning a move to greener pastures. After fetching me another pint, he told me that the decision had been reversed after heated on-site discussions, and that I could have my job back if I wanted it. I accepted the offer and the beer, and everything was back to normal by the following morning, although perhaps certain individuals in the Forest Service were disappointed that they hadn't seen the last of me. The damage to the surrounding timber wasn't too bad, as I recall, and we continued mop-up operations under the watchful eye of a forest officer who had been sent to monitor progress. I seem to recall that he kept his distance.

●

After the Marl Creek upheaval, I tried even harder to figure out the local weather patterns in an attempt to avoid being caught out by these unexpected winds. All the data collected from our company weather stations was great for telling you what had just happened, but what was really needed was a crystal ball for looking into the future, even if it was only a few hours ahead.

As well as the weather stations I'd set up at various locations to gather data for calculating fire danger in the summer and the best time to light up in the burning season, I had rain gauges at our logging camps. They'd be checked daily by the camp managers whenever it rained in that part of the country, as well as by some of the logging crew. The loggers knew that rain during the hottest part of the fire season would lessen the possibility that they'd be put on early shift because of high fire hazard.

One year I noticed some inexplicable rain readings coming from one particular camp that puzzled me for a while, as none of the surrounding area had received a drop. I eventually figured out what was going on: one of the loggers had been pissing in the rain gauge on a regular basis. To counter this, I nailed it a few feet higher up, so he'd be forced to use a stepladder. I couldn't really make a fuss about it, as I'd been doing more or less the same thing to the hazard sticks that the local Forest Service staff had been putting out in the woods near our weather stations.

They'd been doing this in an attempt to prove that we were missing opportunities to burn slash by compiling data from these sticks, each of which was a set of standard wooden dowels that would be weighed to determine moisture content. At that time hazard sticks were considered obsolete by most intelligent people; they'd been replaced by the Canadian Forest Fire Weather Index System, which gave much more scientific and accurate figures for determining fire danger and forest fuel flammability.

Certain members of the Forest Service were trying to catch us out, and they knew that we knew, so they'd hide the hazard sticks where we wouldn't find them. Different people would be sent out to weigh them from time to time to get an estimate of how dry the woods were, and some of them had trouble following directions. To make it easy for them, there'd often be a piece of flagging tape tied to a tree beside the road to mark the start of the path in to the sticks. Once we figured this out, the rest was easy. I'd stop by from time to time and piss on the sticks, and after I mentioned this to my supervisor, he did the same. It must have really puzzled the Forest Service to discover that there'd been heavy localized showers where their sticks were hidden.

These amusements came to an end one day, unfortunately, thanks to the same supervisor. He walked up to me with a big grin on his face when I returned to the office and informed me that he'd taken a dump on one set of hazard sticks. The Forest Service wouldn't need to call in a forensic expert to determine that the deposit on their sticks was of human origin. (They had shit on us often enough, however, so it was probably about time we got even.)

Humans weren't the only ones to interfere with fire weather equipment—porcupines decided to have a go at the weather stations. Probably they wanted a change of menu from their regular diet of plywood road signs, radiator hoses, hydraulic lines, tires and seat cushions on parked logging equipment. I tried various deterrent methods to try and stop them from chewing through the wood into the weather equipment and biting through the wires, but metal sheathing and barbed wire didn't deter them. I even hung up a dead porcupine next to one weather station as a warning, but it only seemed to have an effect on the poor guy who had to change the instrument charts while I was away. He was somewhat disconcerted by the bits of rotting porcupine sloughing off as he worked downwind. The rest of the local porcupine population were not at all put off by the stench.

•

The most useful piece of equipment on our mop-up operations was the Ukrainian Water Bomber, which I'd first encountered on the Double Eddy Creek fire: a log skidder with a water tank attached. On some of our spring burn operations, we'd use several of them as a way of delivering water quickly to wherever it was needed within a burn. The Workers' Compensation Board (WCB) would probably have frowned on the way we used them to carry passengers up steep Cat trails and across the river on occasion when there was no bridge and the water was too deep for wading. Come to think of it, the Ministry of Environment wouldn't have been too thrilled at the latter use either, as sometimes the skidders would be leaking oil and/or hydraulic fluid into the water on the way over. The operators liked these river crossings, as it saved them from having to wash accumulated mud off the machines after they'd been stuck in soft ground.

On those days when tiredness or hangovers made the prospect of hiking up the hill one more time uninviting, a skidder would be flagged down for a lift. We did this one Sunday morning when all of us on the crew, including the skidder operator, were suffering from the combined effect of altitude and the previous night's partying. (The tank of oxygen I carried behind the seat for emergencies was passed around at times like these.) All went well on the way up the hill, until one of the tires suddenly struck a stump and the impact shook most of us off. I landed hard on the ground, accompanied by the case of dynamite I'd been carrying along for opening up an underground fire, followed closely by the box of detonating caps. I'd built it myself and had installed foam padding on the inside to protect the contents from this kind of mishap. My instinctive reaction when I hit the ground had been to throw the box in a different direction from the dynamite, which would reduce the size of the explosion if the caps were to explode. Fortunately, nobody ended up underneath the skidder tires, and the cap box was intact.

●

"Persons who deliberately start fires in our forests are criminals who belong behind bars."
—From a BC Ministry of Forests and Lands leaflet

5
FUELLING THE FIRE

The handheld driptorches we carried for lighting slash used fuel that was a blend of gasoline and diesel. The proportions weren't consistent, as nobody took that much care when the tanks were being filled in the backs of the trucks. At times there'd be too much gasoline in the mix, which made the torches really easy to light, but when the burning fuel was dripped as you walked along, it would burn out too fast, before the slash had a chance to ignite properly. (Now and then someone new to the light-up crew would be handed a torch filled with straight gasoline, and all eyes would be on him as he singed his eyebrows while getting it lit.) Too much diesel in the mix not only made the torches harder to light but made everything from trucks to crew smell even more like a refinery. Sandwiches tasted of diesel regardless, if you'd neglected to wear gloves while handling driptorches. Fuel got spilled on clothing, so you'd sometimes have to be careful about standing next to a burning landing pile to warm up in freezing weather, as there'd be a chance that you'd go up with it.

A certain amount of fuel would end up spilled in the back of the trucks, which made everything carried there smell strongly of diesel. This didn't bother me too much, as my time in the solvents plant had left me quite used to the aroma of hydrocarbons. Others were more susceptible, as I discovered while staying at a hotel in Nakusp late one season. After taking my luggage (a toothbrush and a case of beer) up to my room, I discovered that the beer was unacceptably warm and there was no ice machine available. I improvised by bringing up a bucketful of snow from the back of my truck and dumping it in the wash basin, which did a great job of cooling the beer. Unfortunately I'd forgotten that the snow was soaked with diesel, so that the air in the room was thick with fumes by the next morning.

The fall burning season generally started with broadcast burning in early September and ended with the last landing burns in late November. In later years we'd have to begin high-elevation broadcast burns in late August before the snows arrived, which meant by the time the end of November rolled around, my clothes would have been ideal attire for a Buddhist monk preparing for self-immolation. It wasn't all bad though, as it meant I'd be allowed to go first whenever there was a lineup at the supermarket. On the other hand, when two of us were once in a restaurant in Revelstoke, about to order, the proprietor came over to our table to ask us whether we'd been working with diesel. Apparently the other diners were complaining that the smell was putting them off their food, and we ended up being thrown out of the restaurant.

My first burning jacket was a discard I'd found while rooting through the miscellaneous trash in one of the crummy buses during spring breakup. The back of it ended up being burned through while I was lighting a landing, with the result that the feather stuffing started falling out. The company office staff weren't too thrilled by the fact that I'd leave a trail of dirty feathers behind me when I walked through the building, so I mended the damage with duct tape. Eventually the jacket got so soaked in diesel oil that it was too much even for me. It was ritually burned at the end of one burning season, along with the jersey and jeans I'd been wearing concurrently.

My second burning jacket was an old Ski-Doo jacket my brother had found on a garbage heap out in the woods and handed down to me. It lasted for the rest of the time I spent slashburning, and I have it to this day. Many times in the last few years I've tried to throw it out, as it's literally falling apart and encrusted with years of accumulated filth, but for sentimental reasons I just can't do it.

There was a superstition that it was bad luck to change your clothes during the burning season. I don't know where it originated, but many seemed to take it seriously. For most of us who were single, it wasn't a great hardship, but those of the crew with wives and/or girlfriends ran into opposition. The girlfriend of one member of the crew insisted he leave his burning clothes out on the porch whenever he visited her place. During that season he ended up getting hurt during a mop-up operation when he fell onto a log and managed to get a branch stub driven into his thigh. When I got to him I discovered that there was a small chunk of flesh still adhering to the end of the stub. He thought that perhaps we should poke it back into his thigh, but I didn't think that was a particularly good idea, so we left it where it was. Fortunately he was able to walk—if I'd been forced to pack him out, I'd have had to quarter him first. The doctor who fixed him up back in town said it was the dirtiest puncture wound he'd ever seen, as it contained fragments of diesel-soaked denim. After questioning the patient afterwards, I discovered that he had in fact changed his pants a week or two previously, which just goes to show that you defy superstition at your own peril.

This was the same individual who applied his lighter to the bottom of the newspaper I was reading in our cabin as retaliation for my applying my lighter to the frayed end of his fuel-impregnated jeans a day or two earlier. He'd had to beat out the flames as they climbed up his leg, and I'd been expecting him to get even. As my newspaper burned fiercely, I pretended not to notice and kept on reading until only the edges I was holding remained. I'd sooner have suffered second-degree burns than give him any satisfaction by reacting. None of the crew ever did suffer a burn of that nature all the years we were burning, which is somewhat surprising considering the fact that we worked with volatile fuels and fire on a daily basis.

The risk factor did go up a notch, though, when a flamethrower was added to our burning tools.

After seeing one in use while burning with the Salix crew near Revelstoke, I constructed my own simple flamethrower from a piss can, which only involved adding a check valve and a wick. The check valve wasn't essential for its operation, but it did prevent flame from burning back into the tank and turning it into a firebomb. It was my very own Puff the Magic Dragon. That evening I tried it out on logging slash while we were lighting up a block in the Beaverfoot Valley, and I found it worked really well, sending out a flame that was up to twenty-five feet long. The next test was to see what psychological effect it would have on personnel, which I conducted after it got dark. I hid behind a stump at the side of the road and ambushed what I thought was one of our crew vehicles, firing a burst just inches ahead of the front of the truck. The vehicle immediately slammed on its brakes and the occupants bailed out of both doors, but it turned out to be a Forest Service vehicle, not my crew. I expected to catch shit, but instead they were so impressed by the blast of flame that they wanted me to build them a flamethrower too.

This flamethrower was used for several years in all kinds of conditions. It stood up to abuse remarkably well, as I discovered while burning landing piles late one season. The logs in the landing were covered with snow, which made climbing around very treacherous, since I wasn't wearing caulk boots, which have spiked soles that provide grip when walking on logs. I slipped and slithered my way along, searching for a cavity in the steep face of the landing where there would be drier fuel for lighting. All of a sudden I lost my balance and had to let go of the flamethrower, which fell off the landing pile and ended up battered but intact, with the wick still burning. My partner turned to see what all the cursing was about, and once he spotted the burning wick, he moved hastily further away in case the tank should explode. I wasn't about to lose such a useful tool, so I slowly worked my way down the slippery logs until I was able to retrieve it.

Over time the flamethrower tank received further dents from rattling around in the back of my truck, and leaks became a common problem. The hand pump suffered also, and one evening it nearly resulted in my truck catching fire. I was driving slowly along the road beside a block we were lighting at Smith Creek while one of the crew stood in the back, lighting the slash at the top of the cutbank with the flamethrower. He was unaware that burning fuel was dripping from a loose fitting on the pump and falling onto the side of the truck, where it was running over one of the fuel tank caps. Luckily the thick accumulation of road dust and grime covering the truck protected most of the paintwork.

I even managed to try a bit of hunting with the flamethrower when we were on the way to burn landings in the Beaverfoot Valley. As I turned a corner, I spotted a couple of grouse standing in the middle of the road looking rather edible, but I didn't have a gun with me to pick them off. However, Puff was sitting in the back of the truck as usual, so I ended up creeping along the ditch toward the birds with the wick lit, ready to fire at an instant once I got close enough. My shot fell short, and the birds flew up to a nearby tree to avoid the line of flame on the road and the thick black smoke. Without thinking, I immediately fired a burst into the tree, which again was a fraction short. The birds flew off, but the tree started to burn nicely, to the amusement of the burning crew who had pulled up behind my truck. As it was late October and a seldom-used back road, we decided to let the tree burn itself out, and we drove on.

The next flamethrower I experimented with was a Panama torch filled with napalm and pressurized with nitrogen. A Panama torch is a slashburning tool that uses liquid fuel, which is designed to be pressurized with an integral hand pump and carried on your back. It was never designed for using napalm gel, and the back-check valve may not have worked properly as a result. I figured nitrogen was much safer than compressed air, as it would extinguish any accidental blowback into the tank. I was disappointed with the initial test results, as the tongue of flame didn't reach out as far as I'd hoped. I put more nitrogen into the tank and tried again but still wasn't happy. I kept increasing the pressure in the tank, and eventually I got a nice long flame, but by this time the pressure gauge was well over the red line and the tank was making peculiar squeaking sounds. I didn't relish the thought of carrying it on my back in the slash in this condition, and I suspected none of the crew would either, so I scrapped the project.

The last flamethrower I built was made from an obsolete helitorch and designed to be carried in the back of a pickup truck. I built a handheld gun assembly using bits and pieces scavenged from the scrap metal pile at the back of the mill yard and modified the electrical system so that I'd get a nice fat spark at the igniter tip. Since the helitorch used a gear pump and the fuel was napalm gel, I expected to get a reasonable tongue of flame when I pulled the trigger, which was actually just a toggle switch. I dragged the whole assembly out through the double doors of the fire shed, aimed the gun straight ahead and fired. The result was better than I expected: the gun recoiled as it threw a flame at least eighty feet, narrowly missing the forklift that had unexpectedly appeared around the side of the building just as I fired. It skidded to a stop, and luckily the driver was quite understanding—so much so that I had to demonstrate the flamethrower a couple more times. It was at that point that I realized I'd better add a safety feature to prevent someone from accidentally hitting the switch and incinerating anyone standing in front of the nozzle. Accordingly I wired in a second toggle switch so that you'd have to pull both at the same time to make the flamethrower work. When I eventually left Evans Forest Products, I explained to them exactly how the torch worked and made it clear that I refused to be held responsible for any accidents that might result from its use from that time on.

Soon after I started broadcast burning, we replaced the old liquid fuel helitorch with one that used napalm, which did a much better job of ignition, particularly if the slash was slightly on the damp side. A helitorch is an ignition device that's slung underneath a helicopter and used for lighting logging slash by pumping fuel past an igniter when the pilot turns it on. This newer torch dropped a stream of burning gel that coated everything it landed on as it was flown along. You didn't want to be underneath the stuff—it wasn't easy to beat the flames out, as I'm sure those on the receiving end of it during the Vietnam War would agree. (Some of the powder we purchased for mixing our napalm was actually surplus from that conflict, and it came in drums with military markings.)

Watching the torch in operation was always spectacular, particularly when the helicopter was high up in the air. One evening we watched, spellbound, as the helicopter climbed upward in a spiral over the nearby lake while the last of the napalm was pumped out in a shower of twinkling droplets. We'd encounter problems if the napalm mix was too thin, though, as it would flare up and burn back toward the drum on the torch, which wasn't good. I was sitting next to the pilot when this happened while we were lighting a block on the West Columbia, northwest of Golden, and the flames began to set fire to the torch's electrical wiring. When the pilot pressed the cargo release button to jettison the helitorch, nothing happened, so he pulled the manual release handle. Still nothing happened, and by now the wiring was starting to burn nicely as he flew around in a circle. I eventually got the release to work by pulling on it with both hands, and the torch dropped to the ground, where the crew was waiting to extinguish it.

During one helicopter light-up operation I discovered that one of the crew was mixing napalm in a steel helitorch drum with a steel fire shovel, as they'd left the proper aluminium (non-sparking) paddle back in town. To make it even more interesting, both he and his partner were smoking, and the ash from their cigarettes looked like it was about to fall off into the drum any second. As I backed away, I reminded them of the extremely low flashpoint of gasoline. They assured me they knew what they were doing, and that the ash was cold by the time it fell off anyway. Amazingly, nobody ever blew themselves up during our burning operations, so perhaps the Devil does look after his own.

These days it seems that helitorch fuelling operations are not quite as free and easy, with personnel wearing Nomex flame-resistant coveralls, grounding wires being laid out to prevent static buildup, and fire extinguishers strategically placed about in case of an accident. Our safety procedures were much simpler—there were none.

I'd carry empty helitorch drums in the back of my truck quite often during broadcast season, where they'd bounce around with the rest of the loose equipment. Sometimes they weren't completely empty, with a few inches of napalm remaining at the bottom of the drums, which came in handy, as I could scoop out a cupful of the glutinous stuff whenever I needed to cook something out in the woods. I found other uses for it too. Early one morning I had a flat tire while driving out into the Beaverfoot. It was still dark, so I scooped some napalm out of the drum, dumped it onto the gravel road beside the truck and lit it so I could see what I was doing as I changed the tire. It was freezing cold, so the heat being generated was very welcome also. The first logging truck that approached saw leaping flames by my truck and the driver stopped to see if I was on fire, then once he found out what it was, he radioed the other trucks to let them know everything was okay.

Once I'd put the spare on, I stamped on the burning napalm to extinguish it, but since I was wearing a pair of old shoes with large holes in the soles (and no socks), the stuff got through and I ended up having to jump into the ditch to put out my flaming footwear. There was a layer of ice on top of the water in the ditch and a layer of mud at the bottom, so it wasn't exactly a great start to the day.

After bouncing around in the back of my truck for a few weeks, one of the helitorch drums needed repair, so I took it to the Woodlands welder. I'd done a thorough job of steam cleaning the drum to remove all of the napalm residue, but the welder flat-out refused to weld the drum for fear of explosion. I went over to the plywood plant to ask the welders there if they'd do the work. The first one told me to piss off, as he had the same concerns, but the next one I asked said he was willing to give it a go. I ended up with my head and arm through the hatch on the drum, pushing on the metal from the inside so that he could do the welding on the outside. I can remember hearing his voice echoing inside the drum as he announced, "If it does explode, you should get blown out through that double doorway behind you."

Napalm could even be used for recreational purposes. As were heading out to attempt one last broadcast burn late one season, we found the Trans-Canada Highway blocked by an accident that had taken place west of Quartz Creek, located west of Golden. From the look of the weather it seemed extremely unlikely that we'd be able to carry out the burn, as it was snowing, and there was probably so much snow on the block we were heading to at higher elevation that it would be impossible to light. We stayed in the traffic lineup for a while until it got boring, at which point I dug out an old golf club from the back of the truck and drove a few balls into the forest beside the highway. To make it a bit more interesting, I smeared them with napalm and then lit the golf balls before hitting them into the trees. For some reason all the other people in the traffic lineup seemed a bit nervous, though I can't imagine why, as the woods weren't dry enough to catch fire at that time of the year.

The military surplus napalm thickener was rather unpleasant stuff to work with, as it would cause severe irritation if it got up your nose, and you'd try not to breathe it in. I soon discovered that it was best to stand upwind when measuring out the powder, or better still, get someone else to do the job. (Eventually we switched to a better brand of napalm thickener that wasn't quite so irritating when it was inhaled.)

I was driving out to a burn one afternoon with a drum of the stuff in the back of my pickup and failed to notice that the lid had fallen off. The slight negative pressure in the back of the truck was drawing the powder out and distributing it in my wake. It was starting to rain lightly, which made things really nasty for the car following me. When I looked in the rear-view mirror, I saw the driver peering anxiously through a thin film of what looked like green snot as his windshield wipers tried to deal with the powder that was accumulating on the wet glass. Fortunately my turnoff was just a couple of miles ahead, so I sped up quickly and soon vanished from his sight. To this day he's probably wondering what the hell it was that landed on him.

•

The helitorch didn't break down too often, but when it did, it was always at a critical moment. Once ignition has started on a block it has to be continued in order for the burn to be carried off successfully, particularly in the case of a convection burn. If the helitorch should stop working before the ignition sequence is completed, the central fire can die down and you'll lose the indraft that is necessary to keep the burn contained. This happened to us while we were lighting up a block in the Beaverfoot Valley. The helicopter had made a pass with the helitorch, but something went wrong with it and the pilot had to land to find the problem. I wasn't certain if it could be fixed in time, and darkness was fast approaching, so I went in with the crew to finish lighting the block by hand. We weren't aware that the pilot had fixed the torch until he suddenly flew overhead and started lighting again. It was quite dark by then, so the shower of flaming napalm looked really pretty against the sky. We marvelled at the spectacular sight until we suddenly realized that he couldn't see us on the ground and was heading straight for us, whereupon we had to scramble quickly over the slash to get out of the way.

Now and then the problem connected with the helitorch was pilot error, as was the case during a broadcast burn in the Bush River valley. The pilot had been flying the torch during light-up and had circled around so his flight path took him over another block on the other side of the river. This block had no road access, as the winter bridge across the river had been removed, and it wasn't scheduled for burning. During radio communication with the ground crew, the pilot accidentally pressed the button on his cyclic control stick that turned on the helitorch slung under the helicopter, instead of the microphone button, resulting in a few blobs of burning napalm landing in the slash. He thought it better not to explain what had actually happened, so he merely informed me helpfully that he'd just noticed a spot fire on the block below him. The wind was blowing the convection column above the block we were lighting in the opposite direction from where he was, so it was physically impossible for a spot fire to have started. One of the company foresters and I ended up wading, laden with pump and hose, through the chest-deep, ice-cold water in order to put the fire out. I puzzled over how this fire started for years, until one of the crew told me that the pilot had confessed to him one day.

This wasn't the only time this particular pilot's trigger finger had malfunctioned. During light-up of a block on the West Columbia he'd accidentally flown over one thousand feet on the wrong side of the fireguard. There were too many witnesses to that mishap, so he couldn't very well blame it on sparks blowing across the fireguard. It could be that the copy of the burning plan he had with him had somehow got accidentally turned upside down.

Usually the pilots flew without anyone else on board when the helitorch was in operation, as it reduced the amount of weight the helicopter would have to lift as it flew around the block. Sometimes getting airborne with a heavy load could take a bit of work, as we discovered one day after I'd loaded the cargo compartment and back seat of the helicopter with jerry cans of gasoline and then climbed aboard next to the pilot. The helitorch was attached to the machine as well, and though we managed to lift off the ground, there wasn't enough power to lift the helitorch as well, as it had a full drum of napalm installed. The pilot ended up flying forward a few feet off the ground, dragging the helitorch along the gravel road as he tried to get enough momentum and lift to get everything airborne. Fortunately we didn't seem to generate too many sparks as the torch scraped and bounced along behind us.

I enjoyed the few occasions I was able to fly with pilots when the torch was slung underneath, even if we weren't actually burning slash with it. One afternoon we were flying back to town after completing a burn, ahead of the crew, who were driving back with the rest of the equipment. The road they were on wound through cutblocks separated by residual standing timber, so they'd soon lost sight of the helicopter. We decided to give them a surprise, and the pilot laid a line of burning napalm across the road just ahead of the lead truck. They were turning a corner in the timber at the time and didn't see the flames until the last moment. As they screeched to a halt, the pilot laid another line of fire behind the trucks so they couldn't back up. They weren't able to drive through the flames, as one vehicle had a leaking fuel tank in the back that was leaving a trail of flammable liquid on the road behind. The crew had to wait until the fires burned out, but as they pointed out to me when I met them later in the bar, they didn't mind the holdup in the least, as they were on overtime by then.

We did something similar while flying over a block we were about to burn on Blackwater Ridge so that the pilot and I could finalize the burning sequence. There were two hunters sitting in the block who didn't seem to get the message that we wanted them to clear out, as the whole place would shortly be going up in smoke. The torch was slung underneath, so the pilot hovered over a rock outcrop and dropped some burning napalm. The hunters got the message and took off like scared rabbits. The same tactic was used to chase a moose out of another block nearby a few days later.

One of the best experiences I had was flying in a helicopter over a block after it had been lit up, after dark. The sight of the mass of flames below us, and the heat that was radiating upward, led me to contemplate where I'd most likely end up in an afterlife.

•

Relevant Facts:
- The flashpoint of gasoline is around minus forty degrees Celsius.
- The regrowth rate of scorched eyebrows is approximately 0.16 mm per day.

6
DRIVING WATER UPHILL

Fire has no difficulty climbing forested mountainsides, and if the conditions are right, it will do so very quickly indeed. There's nothing quite like watching a crown fire roar all the way up to the treeline, as long as you're watching from below—or better still, from a good vantage point on the other side of the valley. Water, on the other hand, prefers going downhill, and it takes persuasion to get it to go up a mountainside so that the upper reaches of a wildfire or escaped slashburn can be extinguished. This persuasion comes in the form of high-pressure pumps, the most common one being the Wajax Mark 3. For trivia enthusiasts, the name Wajax was coined by Watson Jack & Co. Ltd. of Montreal, inventors of this centrifugal water pump, which is powered by a Rotax two-cycle engine. I came up with a number of alternative names for the Wajax fire pump over the years, mostly unprintable. The Rotax engine could be temperamental, usually choosing to break down in the most inconvenient situations.

Our company had a total of sixteen Wajax pumps: thirteen of the Mark 3s and three of its little brother, the Mark 26. Every spring we'd check them all over and test them down at the river to make sure they were all working properly, but every year there'd be some that would stubbornly refuse to start when needed urgently on a fire. To compensate for this, we'd usually send two pumps whenever one was requested, in the hope that at least one of them would work.

I eventually figured out that some of the problems were caused by mix oil gumming up the needle valve in the carburetors when residual fuel inside slowly evaporated. From then on I tried to make everyone using the pumps disconnect the fuel hoses when shutting them down rather than just push the stop switches, so that the carburetors ran dry. There's another good reason for this routine when water's being pumped a long way uphill and someone has forgotten to install a back-check valve to prevent water from running all the way back down the hose when the pump is shut off. In this situation, if you don't hold the stop switch down long enough, the engine may restart when you take your thumb off, and then the force of the water pushing backwards can shear the crankshaft. When pump operators continued to use the stop switches, I finally removed them altogether (the switches, not the operators).

I got to know the pumps intimately and had many a frustrating session with them in the field, coaxing them to start, and in the fire shed, taking them apart. I once became so frustrated with a pump that had been having chronic problems that I finally got thoroughly pissed off and sliced the carburetor off the pump with a double-bit axe. I then threw the axe into the wall, just as my long-suffering supervisor happened to be coming through the door. He prudently turned around and left without a word, whereupon I had to remove the remains of the sheared-off bolts and install a new carburetor before I went on to delve further into the original problem.

In the early years (before I really got to know it), I thought the Mark 3 was such a great pump that it deserved the best treatment. Accordingly, when a couple of us were sent out to set one up for supplying water temporarily to the logging camp that was being constructed at Gold River beside Kinbasket Lake, I built a small shelter for it to protect it from the rain. Unfortunately I'd forgotten to cut a big enough hole in the side next to the exhaust, as we discovered when we returned in the afternoon to find the wooden structure starting to burn merrily.

We never burned up any of these pumps on a forest fire, although one did get accidentally dragged behind a skidder for some distance and had a few external parts broken off. The most expensive repair was a result of human error, or rather human stupidity. The Wajax is fitted with an overspeed cut-out device, which is designed to kill the ignition if the engine speed becomes excessive due to the centrifugal pump not getting water. On this occasion we'd been pumping water out of a swamp, and the suction hose strainer was getting clogged with all the organic muck lying on the bottom. The crew boss was tired of having to go and clear the strainer every time the pump quit and had decided to wire down the safety device so it would keep running. By the time I discovered what he'd done, the pump unit was hot enough to fry an egg and was virtually a writeoff. This was the same crew boss who had a day earlier informed me that he'd just seen a Martian spaceship land on the other side of the swamp, so there was obviously something wrong with his grasp of reality.

Wajax pumps ran at high speeds and vibrated quite a lot, which could cause problems whenever one was set up beside a river. The vibration would cause them to gradually bury themselves in the silt as water was brought to the surface, in much the same way as liquefaction occurs during an earthquake. To counter this problem, I attached plywood pads to the bottom of the pump frames, which kept them from becoming mired down but didn't help much whenever the river rose unexpectedly during the day. More than one pump was drowned in this manner, but as they were attached to hose lines, all you had to do was haul them in like hooked fish and dry them out.

Our Wajax pumps weren't the only ones in the fire tool cache; we had others that were even more frustrating: the three Gorman-Rupp pumps. They were complete pieces of shit with major design flaws, such as the fact that the carburetor was mounted upside down, and the muffler was situated right where you'd naturally put your hand to brace yourself while pulling the starting cord. You'd get a nasty burn on your hand whenever you went to restart the pump while the exhaust was still hot. I was very happy when these rotten relics were finally decommissioned, as it saved me from having to take them out in the woods and run a Cat back and forth across them a few times.

Fire pumps can only lift water so high through a hose line stretched up a mountainside, as it takes half a pound of pressure to raise water one foot of elevation. The hose itself reduces the final pressure at the nozzle end due to friction caused by the inside lining of each hundred-foot length, and there's something called the Pump Equation that's used to figure it all out. The rule of thumb, however, is that if you can piss further than the stream emerging from the nozzle, the system's running at less than peak efficiency.

Getting the hose lines laid out up the mountainside usually requires a lot of effort, as there's a limit to how many hundred-foot rolls can be carried by one person. This limit is reduced still further when the ground is so steep and rough that you need one hand to hang onto things on the way up. Sometimes we'd arrange to have rolls of hose dropped onto the mountainside by someone throwing them out the door of a helicopter, but it wasn't always successful at first. Some of the hose would roll off downhill, and one or two lengths never were found, so I came up with a modified method: two rolls of hose would be connected together before they were tossed out. This meant that if a roll decided to race off down the mountain, the one it was paired with would hopefully serve as an anchor.

It was very important when pumping water up a steep slope that the hose be tied to a secure anchor point such as a tree or a stump, because once the pump was started, the weight of water within the hose would pull it down. I saw this happen one day when the crew was working on an escape above a steep cutblock at the back of the Bush River valley. They'd been reminded to tie off the hose, but they replied that they knew what they were doing, and did I think they were idiots? That question was answered shortly after the pump was started, as several hundred feet of hose slithered straight back down the slope like a giant strand of spaghetti. I left them to sort it all out.

Forestry hose wasn't cheap: in those days it was over a dollar a foot, so the idea was to avoid burning it up if at all possible. Sometimes the loss was unavoidable, as was the case at Moose Creek one spring when an escaped burn we were working on took off suddenly due to strong winds and chased us out. I told the crew I was with at the time to head to safety, and asked them if they'd mind saving as much hose as they had time to uncouple and drag with them on the way out. Unfortunately I couldn't follow them, as I had to head off in the other direction to round up a two-man crew whose escape route had been cut off, and in the end none of the hose was rescued. I went back later to discover some nice straight lines of carbon marked with metal couplings every hundred feet.

Now and then hose was lost through sheer stupidity, which happened one day when someone dropped four brand-new rolls of hose onto a hot spot within a burn we were mopping up. By the time I found it there were so many holes burned through the hose that it was all a complete writeoff.

The company's supply of hose wasn't unlimited, and once the cache was depleted, I'd have to repair hose that had become damaged and get it back into service as quickly as possible. This would mean that I'd have to drive back to the fire shed in town at night, mend those hoses that weren't a complete writeoff, then drive back out to the woods. Sometimes I wouldn't get back to camp until the small hours of the morning.

Pulling hoses off a fire after it was out could be a miserable job, particularly if it was raining. For some reason the couplings trailing behind seemed to frequently get snagged on obstacles whenever you were ascending or descending steep ground, so you'd constantly be going back to free them. It didn't seem to happen nearly as often on flat ground, for some reason... Sometimes the crew would get a Cat or skidder to drag the hose uphill, with them hanging onto the end for a free tow, which would be easier on their backs but frequently hard on the hoses. Once or twice a firehose was used as a tow rope when a truck had a flat battery, and that wouldn't do it much good either.

•

Sometimes it was possible to take advantage of gravity to bring water to a fire, as we did on the Smith Creek fire mentioned earlier. A gravity supply would run twenty-four hours a day and didn't require that pumps and fuel be lugged up mountainsides. The only problem might be silt building up where there were low points on the hose line. With enough of a drop in elevation, it's possible to get good pressure at the nozzle. It's a miniature version of the hydraulic sluicing system used in placer gold-mining operations.

I managed to discover a way to screw up a gravity water supply when I was setting up a source of drinking water for a trailer that had been placed on site in the woods for the operator of a portable sawmill. I ran a hose line up the creek far enough to get reasonable water flow, then attached an intake cone (known as a gravity sock) to the end and set it in the creek. When I finally returned to the trailer and opened the door, I found the floor was an inch deep in water, as the holding tank that was built in under the bed had overflowed. It had never occurred to me to install a tap at the lower end of the line. At least all the accumulated cigarette butts got washed out the door.

Whenever there wasn't a river, creek or other water source suitable for a pump set-up near a fire, we'd have to haul water in by tanker truck. This could be quite pleasant work on a warm day, as you could lie on top of the tank and doze in the sun while it was being filled up at a creek, with one hand just inside the hatch so that you'd get a warning before it overflowed and washed you off. Another convenient source of water for the tanker would be the small streams that flowed through culverts under the logging roads. We'd block off the upstream side of the culvert with a piece of plywood to dam the water up temporarily so it would be deep enough for the drafting pump. This was a great set-up, except for the odd time that we forgot to pull the plywood back out before driving off, and the water rose to the point that the road washed out.

It was while filling a tanker on a frosty morning that I came up with a brilliant idea for getting warm. While standing on top of the pump motor housing with my back to the exhaust stack, I held my jacket over it and let the hot air flow up around my body. It wasn't until I started feeling dizzy that I realized I was being saturated with carbon monoxide. The pounding headache I carried around for the rest of the morning served to remind me that it hadn't been such a great idea after all, even if the fumes had killed off any minute insect life living in my jacket.

Sometimes later in the burning season it would get so cold that the pump on the tanker would freeze overnight. When this happened, the cure was to shovel hot coals from the nearby fire onto the steel deck under the pump to thaw it out. The hoses would have ice in them, and it would take some time to blow it all out before we could get water flowing through again. We had the same problem when conducting spring burning operations at high elevations.

The strangest situation occurred in November one year when a fire started in the middle of felled timber on a logging operation up in Copper Creek. The temperature was about minus twenty Celsius, so naturally all the creeks on the mountainside were frozen solid. The tankers had been put away for the year, so two of us headed out with a few piss cans full of water to which we'd added some antifreeze. Not enough as it turned out: by the time we arrived on site, we found the contents of the piss cans were also frozen solid. It didn't really matter in the end, as the felled timber was sitting on top of a couple of feet of snow, so once the burning material was isolated from the rest of the fuel, it went out on its own.

Our tankers had other uses beyond hauling water to fires and serving as sunbathing platforms. They made great mobile coolers if you took the insides out of an industrial engine oil filter and hung it inside the tank on a piece of rope secured to the hatch lid. It was possible to cool almost a dozen beer at once with this set-up, as long as the rope didn't chafe through along the uneven logging roads.

You didn't want to drink water out of a tanker, however. Ours weren't too bad, as the only contaminants were rust and residual oil from our improvised coolers, unless they'd been filled up from some particularly stagnant swamp. Hired tankers being used on a fire were less safe, as there was always the possibility that they might have recently been used for hauling liquid sewage.

Even though I knew our company tankers had never been used for hauling sewage (apart from the odd occasion when one of the crew might have pissed into the tank), I never drank out of them—particularly after the pump on the Kenworth tanker had an oil-injection system added to improve the lifting capacity, which had been reduced by constant wear from abrasive silt in the water. My answer to the problem seemed to be a good one until I noticed that every time I pumped out of a lake, there'd be a significant oil slick developing as some oil leaked back out. This was not the kind of thing to endear us to the Ministry of Environment, so I looked through the plywood plant to find a substitute for motor oil and discovered a couple of drums of polyethylene glycol 2000. I'm not sure if the aquatic life enjoyed the taste any more than that of the oil, but at least we weren't leaving visible evidence anymore, as this stuff conveniently dissolved in water.

•

I eventually took an air brake course and obtained a Class 3 driver's licence so that I could legally drive the company tankers on the highway. Prior to this I'd already been driving them on the logging roads, as in those days nobody was too concerned about whether you had the correct licence as long as you stayed on the back roads. I wasn't exactly a model driver—I tended to drive the tankers the way I'd drive a pickup truck. Unfortunately, with a full load of water they wouldn't corner on gravel quite as well as pickups, but at least I never managed to roll one over. I had a bad habit of changing gear while going downhill with a load, and an equally bad habit of setting the throttle lock while the tanker was grinding slowly up a long grade so that I could concentrate on opening a can of beans or beef stew as I was driving. These snacks would have been warmed up earlier by being wedged against the engine exhaust manifold.

    The tires on the tankers were usually in terrible shape, as it seems we got the hand-me-downs from other trucks in the company fleet since the tankers were only used for part of the year. Some tires were so worn out that the canvas was showing, and we'd have frequent flats and blowouts. One of them happened as I was driving along the Trans-Canada Highway just ahead of a company pickup truck. I got a call on the radio from the driver following me, informing me that one of the tanker's rear tires was disintegrating and that bits of it were flying all over the place. When I looked in the rear-view mirror to verify what was happening, I could see chunks of rubber flying into the air above the truck and landing on the highway. It got so bad one year that we were getting a flat tire every day, which was a real pain for us but good business for the local tire shop.

    The tankers' air brakes got a good workout on some of the steep grades back in the woods, particularly in the spring when the roads were often in rough shape. I ventured into the Glenogle Valley in the Kenworth tanker one year just after the snow had melted on the roads but before the grader had been in to clean them up. I made it across the first bridge and got partway up a steep pitch, only to find the road blocked by rocks and debris too big to move out of the way. I ended up having to back all the way down around several corners, but as the engine was idling the whole time, the air pressure wasn't building back up in the tanks as I was applying the brakes. For some reason the brakes didn't dynamite when the pressure dropped below the critical level, so I ended up running backwards over the bridge faster than I would have liked, with no brakes at all.

This wasn't the only time I'd had a bit of brake difficulty. While driving the Kenworth down a steep grade on the West Columbia, I discovered that the engine brake (Jake brake) wasn't working, and neither were the air brakes. When I applied the parking brake in an attempt to stop the truck, that didn't work either. By this time I was picking up speed, and I thought it was time to call on the radio and warn upcoming traffic that I was having problems and might have to roll the truck over in the ditch in order to stop it. When I reached for the microphone, I discovered that someone had reached in through the window overnight and stolen it. I finally figured out why the Jake wasn't working and managed to gear down and stop the truck just before it would have gone off the switchback into a steep creek gully.

Odd things happened on flat ground as well. While I was driving the same tanker along a partly overgrown road in the Beaverfoot, a stick suddenly poked through the no-draft window on the driver's side and hit me in the chest right by my heart. I wasn't going very fast, so I was able to stop before it did any serious damage to my shirt.

Like every other piece of equipment in the woods, the old Kenworth had its share of breakdowns. One of the more interesting ones happened out on the West Columbia when the back deck fell off as I was driving along the main logging road. It was far too heavy to pull off to the side of the road, as the deck was made of quarter-inch steel plate and had on it a toolbox and a hose reel holding a couple of thousand feet of wet one-and-a-half-inch firehose. Fortunately a skidder happened to come along the road not long afterwards, so the deck was dragged off to the side of the road, out of the way of traffic. I continued on to town and left the tanker with the mechanics, telling them I'd bring in the rest of the truck the following day.

The next morning I went back out with the warehouse flat-deck truck and got the remains loaded on, with the help of a log loader that happened to be passing by. The reason the deck had fallen off was due to an attempt we'd made a couple of days earlier to split a log the tanker had been backed onto—using Primacord. (We didn't have a chainsaw to cut the log, so I thought I'd improvise.) The explosion cracked some welds, so for the trip back to town we'd tied up that side of the deck with rope in an attempt to lessen the strain on the damaged seams. I didn't get more than five miles before the whole lot fell off. Understandably I never told any of this to the mechanics, as I already had a reputation for being hard on trucks.

Once the back deck had been delivered to the shop for reattachment surgery, I thought that would be the end of the matter. But the next night I received a call on the radio from one of our camps, informing me that they could see a fire on the hillside, suspiciously close to where the tanker had been stuck on that log. I headed out to take a look and sure enough, the Primacord explosion had ignited some rotten wood in the centre of the log, which had smouldered for a few days and then flared to life. No damage was done to the nearby forest, so I didn't think it necessary to bother the Forest Service with a fire report.

In addition to the old Kenworth, the company had a Ford tanker, which had the same twenty-five-hundred-gallon capacity but was a bit underpowered. This would become painfully evident whenever we had to haul a full tank of water up a particularly steep grade on a fire, as the front wheels would lift right off the ground, which was rather disconcerting for the driver and made it a bit difficult to steer properly.

Breakdowns could be very inconvenient, particularly when they happened during a fire situation. When a small fire started on a block that some tree planters were working on in the Beaverfoot, the Kenworth tanker was dispatched. It managed to break down about five miles east of town, so the Ford tanker was sent in its place. The Ford turned out to have a small problem with the air brakes that prevented it from moving at first, but eventually it got rolling out of the yard. Unfortunately its troubles weren't over, as it didn't even get as far as the first tanker before it broke down at the side of the road only two miles east of town. We ended up responding to the fire in a pickup truck with a couple of shovels and a piss can. The fire wasn't very big and had quite possibly been started by a tree planter discarding the remains of whatever he or she had been smoking. Since they'd managed to put most of it out by the time we got there, it wasn't a big issue.

●

There were a couple of others on the crew who would drive tankers during the burning season. They were better drivers than I was and could change gears without making the kind of noises I'd generate when I was wrestling with the Kenworth's split-range transmission. One particular driver was not very good at housekeeping, however, and the cab of whichever tanker he happened to be driving soon turned into a garbage pit. There would be decaying food on the floor, which would eventually work its way under the seat, and all kinds of other filth, including the printed kind, which we'd salvage from the camp bunkhouses for lunchtime reading. He also had a habit of reading newspapers while sitting in the tanker and throwing each page out the window once he'd finished with it. I could always tell which landing he'd been parked on when I was flying overhead in the helicopter, as there'd be newsprint scattered all over it.

This driver was quite large, and he found the cab of the Kenworth to be a little cramped. Often he'd open the door when he was driving in order to make more room for his left foot. The only time I found this to be a problem was when he was backing up in the dark while we were working late on a slashburn. I was crouched on the fuel tank beside the driver's door, guiding him to make sure he didn't go into the ditch, when without thinking he opened the door to get more room, and I was swept off onto the dirt road. I had to roll quickly into the ditch to avoid being run over by the left front wheel.

The lack of room in the Kenworth's cab had caused a problem for another driver a year or two earlier. I happened to be driving behind him when he suddenly skidded to a stop and bailed out the door. As I pulled up behind him to see what the problem was, I found him leaning on the side of the tanker, coughing and retching as a cloud of fine white powder drifted out the open door. It turned out he'd accidentally set off the dry-chemical fire extinguisher while changing gears and had received much of the charge in his face. He recovered after a while and we carried on, but not before the empty extinguisher was hurled into the bushes.

Bush River seemed to be particularly hard on the poor old Kenworth. It got to the point where I started to think there was something wrong about that valley, after all the other problems we'd had on our spring burns. While being driven up a particularly nasty section of road, the tanker ran over a big rock, which put a long crack in the engine oil pan. The mechanics were too busy in town to come out and fix it, so we had a go ourselves. We cleaned the area around the split and applied sealant, but it was too cold at that high elevation for it to set properly, so we then applied heat. In other words, we poured fuel into an empty can, lit it and pushed the can under our repair.

When flames started coming out all around the engine, it became obvious that all the accumulation of oil and grease underneath had caught fire.

Not long after that event the same tanker broke down again; I suspected the trouble was a defective fuel pump. The company decided it was too expensive to fix the tanker, and that we should leave it where it was and hire another one from a local contractor in the meantime. This didn't seem to make a lot of economic sense, but orders were orders. Before we could get the hired tanker past the Kenworth, we had to pump air into the latter's tank in order to release the air brakes so we could drag it to the side of the road. There it sat for several weeks, until I finally managed to persuade the company that it should be retrieved before the snow came; otherwise it would be there all winter. They sent out a mechanic, who confirmed that the problem was indeed the fuel pump, and at last the tanker was put back in service. I had the pleasant task of delivering the bill for the hired tanker to the accounts payable department to be added to all the other invoices relating to the Kenworth.

I ended up spending the night in the cab of the Kenworth while it was parked on a high-lead block in upper Bush River that we'd lit up that evening. The top section had been ignited by hand and was slowly burning downhill. The spur road the tanker was parked on didn't extend all the way to the edge of the block, and I wanted to make sure the fire didn't burn below this level and work its way around to the landings below the road. It wasn't a comfortable night, as the cab hadn't been designed for sleeping in, and there was debris rolling down the hillside onto the road and across the road—some pieces were at least three feet in diameter. When daylight finally arrived, I discovered that a large rock had rolled to within a couple of feet of the tanker before it had come to a stop. But at least the fire hadn't worked its way into the landings.

•

Tankers enabled us to get right up close to the fire whenever one was burning next to a road. When a slashburn jumped across a spur road on Blackwater Ridge one afternoon, I set three nozzles at different angles and then drove through the flames with the windows rolled up, spraying water into the escape. As the pump outlets were only on one side, I had to drive back through the fire, then turn around and repeat the process. After a couple of trips, I got the fire under control with minimal damage to the paintwork.

I nearly wasn't so lucky when I tried to drive through a block at Palliser while it was still burning. It was after dark, and I'd remained behind to keep an eye on things after the crew had gone back to town. This was the same block the tourists had taken pictures of on their way past during Expo 86, and I was taking care that it didn't get away overnight and give passersby some really spectacular shots the following day. I'd noticed that there was a spot fire in the plantation on the hillside above the block, and I decided to try and drive the tanker up to it through the burning block. I got partway through, then found the road blocked by a burning log that had rolled down off a landing pile. It wasn't possible to back up, as the road behind me had a bend and I was unable to see properly due to the intense light from the fire and the contrasting darkness where the road disappeared into the timber. The radiant heat coming off the burning landing pile made it impossible to work outside the cab for more than a few seconds at a time as I tried to drag the log out of the way. Each time I jumped back into the cab to recover from the heat, I'd find my jacket was smoking, and before long I could smell the paint on the side of the tanker starting to soften. Eventually I managed to move the log far enough that I could ram it with a front wheel and carry on driving through the block. It didn't take more than an hour to extinguish the spot fire, and afterwards I curled up in the ditch and slept till dawn.

•

The company finally sold the underpowered Ford tanker to a water-hauling contractor. The next time I saw it was when it was being used on a Forest Service slashburn that I was visiting as the guest of a forest officer. The tanker was apparently on its way down the road toward us, but when we got to where it should have been, all we saw at first was the driver standing in the middle of the logging road. He was speechless and was pointing his finger over the bank. Sure enough, there was our old tanker, upside down, with a couple of its wheels still spinning. It appeared that the driver had pulled over to the side of the road to let us past, when the ground suddenly gave way. He was lucky to have escaped without a scratch, as the tank had prevented the cab from being crushed by the stumps on the slope below the road. I suggested this might be an ideal time to grease the truck undercarriage, as it was now on top, but he didn't think that was very funny. As if his luck wasn't bad enough, he'd forgotten to bring his driver's licence with him that day, as the RCMP discovered when they showed up uninvited.

One morning I received a phone call from the RCMP asking me to bring the company tanker out to a trailer park where a truck was on fire, as the local fire department didn't attend fires outside the town boundaries. It was early in the spring and the tanker had no water in it, so I was able to get to the fire in record time. There were two large fire extinguishers on board that normally would have been adequate to put out a vehicle fire, but this one was different. When I arrived I found a pickup truck blazing, with a plastic kayak on the roof rack that had melted and filled the back of the truck with burning molten plastic. The extinguishers proved useless, as the metal in the truck was so hot that the plastic kept reigniting, and the RCMP officer who'd turned up was getting very worried that the vehicle's fuel tank would explode. He finally shouted that we should all get clear, but I wasn't about to give up, as I believe vehicles only blow up spectacularly in Hollywood movies. I ended up shovelling dirt onto the bubbling mess in the back of the truck until it finally went out.

•

We started using ex-military six-wheel-drive trucks with tanks mounted on the back to haul water up trails that were too steep or muddy for the big tankers. They were fairly successful, with only one of them rolling over. It was put back on its wheels and the thousand-gallon tank was lowered back on again, but the cab was in pretty rough shape. Another of these trucks was being moved from Blackwater Ridge to a new burn site on the West Columbia when I heard a gunshot from across the lake at that approximate location and called on the radio to find out what had happened. I was told that someone in the crewcab had just shot a battery he'd spotted at the side of the road, using a twelve-gauge shotgun. It was discovered a bit later on that the battery had in fact just fallen out of the six-by-six tanker that was up ahead of them.

The good old Ukrainian Water Bomber was used when the skid trails were too steep even for the six-by-six tankers. I got quite used to riding along perched on the tank at the back, though there were risks. During one such trip I saw that we were about to bounce over a stump, and just before the impact I moved my hand from where it had been gripping the steel cable attaching the pump to the tank. Good thing I did, as the cable snapped when the skidder bounced hard, and one frayed end just missed taking off a few of my fingers.

Another skidder that had similarly been set up with a water tank became stuck during a mop-up operation near Moose Creek after going down a skid trail it had been ordered to stay off, as the ground was too soft. The Cat we were using to build fireguard had a ripper on the back instead of a winch, so getting it to pull the skidder out was a bit tricky, as we didn't want to get the Cat stuck as well. The skidder's winch line was securing the five-hundred-gallon tank on the back, so it couldn't be used to help extricate the machine. Several steel chokers were hooked together and then used to connect the back of the skidder to a ripper tooth on the Cat. One member of the mop-up crew was, for some unknown reason, standing on the tank behind the skidder as the Cat started to pull, and the instant I spotted him, I yelled at him to jump off. He just looked at me, so I yelled again and pointed to the ground, at which point he finally jumped, seconds before one of the chokers broke and the severed end came whipping back over the tank. I went over to where the guy was standing in the mud and explained that if he'd stayed where he was, he'd probably be lying in the mud minus a head.

●

"A pump can be ruined in minutes if proper operating procedures are not followed."
—From a BC Ministry of Forests handbook

7
TRACKS IN THE MUD

The four traditional elements—air, fire, earth and water—are of significant importance when it comes to fire control and slashburning. Air, in the form of wind applied to a fire, can have disastrous consequences. Water, when added to earth, creates mud, which can cause problems when heavy machinery encounters it. Fortunately mud isn't encountered too often during firefighting operations in the dry summer months, so wheeled equipment rarely gets mired down. At other times of the year, equipment that runs on steel tracks is more appropriate for the ground conditions. This is when bulldozers become invaluable.

The bulldozer, or Cat, is an ingenious invention, as it can climb steep slopes and go across soft ground that would defeat any wheeled vehicle. They're not invincible, though, and when I started working in the woods, I soon discovered that when one gets stuck, it's *really* stuck. Sometimes it can take hours to get one out, particularly if there aren't any stumps close by to enable it to winch itself out. This was the case when a large Cat that was building fireguard on a steep block at Hope Creek got into difficulties while I was keeping an eye on it. The few stumps that were within reach of the winch cable were quite inadequate and popped out like corks as the operator tried to extricate himself using a combination of winch and reverse gear on the tracks, leaving the Cat bogged down deeper in the mud.

It took us several hours to get the machine unstuck, particularly since we didn't have a chainsaw to cut pieces of log to place under the tracks. We had to scour the hillside for chunks of wood that were big enough for the job but light enough to be dragged down to the mired machine. The operator would push the blade down so that the body of the Cat was lifted off the ground, and we'd crawl under the tracks and fit the logs into place. That wasn't a lot of fun, what with the mud dripping off the tracks that were suspended ominously above our backs and the swarm of blackflies waiting for us to emerge. They didn't seem to mind the language we were using each time the blade was raised and the body of the machine sank back into the mud, taking all our hard work with it, but they probably couldn't hear it over the noise of the engine. Each time I was underneath those tracks I'd wonder what would happen if the hydraulics suddenly let go, but eventually I decided that perhaps being crushed by thirty-five tons of metal might be preferable to being eaten alive by bloodsucking insects.

Getting a bulldozer unstuck from mud was a tricky business, but some operators were masters of the art. Ron was the best, having had much practice operating his Cat in the West Columbia loonshit. That stuff's like muskeg, and once the machine's tracks break through the surface and start digging down, you'd swear it's bottomless.

Ron's Cat became stuck while he was doing some work at the top end of Quartz Creek on a Friday afternoon just before the start of a long weekend. Not that the time and date made the mud any deeper, but anyone working nearby was probably headed for town to start the weekend, with the volume of their two-way radio turned down so they wouldn't have to listen to tales of woe over the air. (If there should be an emergency situation, of course, they'd respond at once. But what classified as an emergency at the start of a long weekend was debatable.) We worked on getting the Cat out of its predicament, using both the main winch and the pony winch, but kept uprooting all the stumps I set chokers on so that the winch lines could be attached. One line snapped as I was inside the cab of the machine and nearly took off a few of my fingers. I'd been gripping the heavy steel grille at the back of the protective cage for support when I decided to change position, and the moment I took my hand away, the end of the severed steel cable whipped back and smacked against the outside of the cage. We were finally down to the last two stumps within reach of the winch lines, and I was thinking we'd have to get another machine sent in to pull us out, when suddenly we began to emerge from the primeval ooze. We barely made it, and I noticed how one of the stumps we were attached to was being lifted half out of the ground as Ron pulled on the pony winch at the same time as he operated the main winch with his other hand and worked the throttle with his foot. It was a masterly performance.

I'm eternally grateful for how he saved my neck the day we were fixing a road washout that had been caused by a blocked culvert. I was standing in the water, clearing debris at the intake end, when the culvert suddenly started flowing and the suction pulled me toward the jagged steel opening. Ron saw what had happened and instantly swung the Cat blade toward me so I'd have something to grab onto before I was sucked in and probably decapitated on the way through the pipe.

After getting a Cat stuck one too many times in the West Columbia loonshit, I decided it would be best to have them working in pairs, as this really wasn't any more expensive than having a single machine get stuck and then being forced to haul in another to pull it out. This worked well, except when both machines got stuck at the same time, which happened when we were constructing fireguards around a wildfire near Double Eddy Creek that had been started by a lightning strike. The fire was burning in logging slash and had been hit with a load of fire retardant from an air tanker, which often doesn't do a lot of good in heavy slash, as the fire can simply burn underneath the retardant coating and pop up again when it gets beyond it. This fire had done exactly that and was heading toward the D8, which was the first machine to get stuck.

The other machine was a D7 and was in the process of assisting its bigger brother when it too got stuck. The fire was getting closer all the time, and as rescue efforts grew more urgent, I was trying to estimate the extent of the insurance claim should both machines not make it out in time. They finally got themselves extricated after a lot of track churning, winching and cursing, which saved a lot of grief and paperwork.

Sometimes a bulldozer would get stuck in the most unexpected places, as I discovered when I got a radio call informing me that a machine working at the top of Goodfellow Creek had fallen into a hole that had suddenly opened up in the middle of a large landing. There were no trees or stumps anywhere within reach of the winch cable, and I ended up sending another Cat on a lowbed all the way out to the site. Once it was there and unloaded, the rescue only took a few minutes, but the bill for the lowbed move was substantial, as it was a long way there and back.

Whenever a Cat got stuck while it was working for me, I would sign for the time it took getting itself out, unless the operator got stuck through his own stupidity or deliberately ignored my instructions. This only happened once, to an operator who was sent to build fireguard for me against my protests. I didn't want to use him, as I knew he was overconfident in his ability to build guard on steep ground, and on site he refused to follow the route I'd laid out on the cutblock, with the result that he spent a few hours wallowing in the mud. It didn't make any difference in the end as it turned out, as he was a good friend of one of the company staff, who overruled my decision to reduce the timesheet by the hours the Cat was stuck.

Once or twice, when the operator wasn't around, I had to borrow a parked Cat or skidder to fix up access to a mixing site on a block we were going to burn. The owners didn't mind as long as they weren't rolled over and they were put back when I'd finished with them. Usually the keys were left in the machines when they were parked, but it didn't matter too much if they weren't, as someone had kindly given me a key that would work on all the bulldozers and skidders made by the Caterpillar company.

From time to time a Cat would have to take a shortcut across a river, which left it clean and sparkling as it emerged at the other side but probably didn't enhance our standing with the Ministry of Environment. Sometimes it wasn't possible to get across a river, as was the case when we tried to get a Cat to the other side of the Incomappleux River (locally known as Fish River) south of Glacier National Park. We were unable to determine how deep the water was due to all the suspended silt, but the operator figured it would be safe to cross. I crossed my fingers as he started in and watched as the front end suddenly plunged into deep water and the machine was enveloped in a cloud of steam. The operator barely managed to back out in time, and we decided to wait until the river level dropped before trying again.

Things didn't work out quite so well when the company D7 Cat ended up in Kinbasket Lake while I was working at Tsar Creek. The machine had been rebuilding the ramp used for unloading trailers for the logging camp we were setting up. The lake level was slowly rising, and the Cat had accidentally fallen in. This was the same machine I'd been borrowing on weekends when the operator had gone back to town so that I could teach myself how to drive a bulldozer. Now it was in a real predicament—only part of the cab and winch were showing above the muddy water, and there was no other Cat within miles we could use to pull it out. Eventually a D8 was sent up by barge, along with a scuba diver, and the drowned Cat was retrieved. The scuba diver was needed to attach the tow cable and release the brakes that had been thoughtfully applied by the operator as the machine headed for its plunge. Somewhat to everyone's amazement, the latter hadn't even got his feet wet when he abandoned his machine as it slid into the lake.

•

Moving loaded logging trucks across bridges could also lead to sudden immersion if the bridge was past the end of its working life. This happened to a truck crossing a bridge over the Blaeberry River when the spruce stringers (main support logs) suddenly snapped under the weight, dropping the truck with its driver and logs into the water. The driver swam to shore, and the truck was salvaged, but the accident led to closer inspection of all the other bridges. Fir logs were stronger than spruce but were more valuable when sent to the mill, so they weren't always wasted on bridge construction. Other species like pine were even less strong and generally weren't used, with one notable exception.

One day I was watching a loaded logging truck drive over a bridge across the Fish River while standing beside a forest officer and the logging contractor who'd built the bridge. This bridge had been constructed with cottonwood stringers, and as the driver gingerly drove across, the whole structure slowly sagged until it was almost touching the water. The forest officer and I watched in horrid fascination, expecting the stringers to snap and drop the truck into the river at any moment, but it made it across safely and the bridge sprang back to its original profile. When it was suggested that perhaps coniferous logs would make stronger and safer stringers, the contractor explained that cottonwood had properties that made it ideal for bridges—it was supple, much the way an archer's bow was. In the logging business, an ideal temporary bridge is one that's cheap to build and lasts until five minutes after the last load of logs is hauled across and it's no longer needed, whereupon it will self-destruct into the river and save the time and effort of removal (unless the stringers are worth salvaging).

Our tankers managed to get stuck from time to time, usually when they had a full load of water, and often in the worst possible location. The Kenworth got bogged down while turning on a narrow road in pouring rain one day, and it was impossible to tow it out, as it was sitting crosswise on the road. After much wasted effort, I ended up driving to town to borrow a heavy chain hoist from the millwrights, and with the help of it and two come-alongs attached to nearby trees, we finally got the tanker unstuck.

The situation was a little more precarious when the same tanker became stuck while it was being filled up next to a small lake. It turned out the ground was softer than it appeared, and as the tank filled up, the tanker started sinking. Unfortunately only the wheels on one side sank in, so the whole machine began to tip sideways, to the point where it looked like it might fall on its side. We ended up draining the water back into the lake and then getting a tractor from a nearby farm to drag the tanker out of its predicament.

Pickup trucks got stuck in the mud more often; at least, it seemed mine did. An example was when I was backing down a nasty piece of road inside a cutblock at the top of Symond Creek in the Beaverfoot Valley. There'd been a recent snowfall, which had started to melt, and this added to the natural slipperiness of the local mud that constituted the road surface. My truck ended up sliding sideways, coming to rest at a steep angle, and I ended up exiting through the passenger door. I was unable to trip the radio repeater from where I was, so I couldn't call for someone to come and tow me out. Eventually I started walking out, and I trudged along the roads for eight miles before being picked up by one of the Woodlands staff who was out doing a little hunting. As the game season had just started, I hadn't been inclined to take shortcuts through cutblocks, as there were generally a few trigger-happy idiots out driving around the Beaverfoot who wouldn't know the difference between a human and a mule deer by the time they'd got halfway through their bottle of Jack Daniel's. I drove back up with the crew the next day to rescue my truck, but it wasn't long before a second pickup slid back to join the first in the mud. Fortunately we had a third vehicle with us and eventually managed to get everything pulled back onto the road again.

Sometimes the mud was even deeper and stickier, as it was where I got stuck on a bad section of road in upper Quartz Creek. Two other pickup trucks also got stuck in the process of trying to extricate me. We finally had to get a bulldozer that was constructing fireguard nearby to come and tow us all out, strung together like a charm bracelet.

You knew you were in trouble when the truck was so deep in the mud that you couldn't get the doors open and would have to climb out a window. Fortunately by the time summer arrived, mud would be replaced by dust, which, while annoying, was a lot easier to live with. It probably didn't do truck engines much good, particularly if their air cleaners weren't changed regularly. When it got really bad, they'd need to be shaken out every day, but not many drivers were that diligent. When the nut and end plate for the air cleaner on the Kenworth tanker fell off onto the road, the driver didn't see the need to do anything about it and continued driving in the clouds of dust. I ended up fabricating a temporary replacement plate from the end of an empty juice can and fixing it in position with a bit of wire. This enabled us to carry on for another few days until it too fell off.

Following another vehicle that was sending up a thick plume of dust could be unpleasant on a blazing-hot day, as you'd have to keep the windows closed, and the inside of the truck would soon reach near-sauna temperatures. It could also be most disconcerting if it turned out to be a loaded logging truck up ahead, as if you got too close there'd suddenly be logs appearing in the murk uncomfortably close to your windshield. Most of the time the logging trucks would pull over and let faster vehicles like pickups overtake them, unless for some reason they didn't particularly like the person following.

The roads out in the woods were hard on pickup trucks, particularly since we'd often be driving up those that had been inactive for years when we were burning old landing piles. Washouts, rocks, logs and deep potholes all took a toll on vehicles, even though they were four-wheel-drive, and according to the claims of one Detroit manufacturer, "built tough." A couple of weeks after I was assigned a brand new three-quarter-ton pickup, the clutch linkage fell apart while I was out in the woods. That was easy enough to fix with a paperclip I found lying on the floor of the truck, but it seemed an odd thing to happen to a truck that had supposedly been through a pre-delivery inspection quite recently. Other problems occurred now and then on that same truck. When I was driving out in the Beaverfoot late at night in a heavy downpour on my way to check for road washouts, the accelerator pedal suddenly went straight to the floor of its own accord, and the truck began to speed up rapidly. I couldn't free the pedal with my foot, and I ended up switching off the ignition, which immediately affected the brakes and steering. Once I'd stopped, I lifted the hood and poked around to see if I could figure out what the problem was but soon gave up, as it was pitch dark and the rain was coming down in buckets. I slept in the truck until dawn, then had another go, hammering on various bits and pieces near the carburetor until I got things working again. I never did find out what had caused the problem, which never happened again, but from the sound of it, certain other vehicle models have experienced similar malfunctions.

I put twelve thousand hard miles on this truck in the first four months, since I tended to drive fast on the logging roads, where speed limits were unheard of in those days. I don't think anyone ever beat my speed record on the A road in the Beaverfoot, and probably the same applies to the B road leading to Bush Harbour. After a while you get to know every pothole and which corners you can take at high speed with minimal sideways drift on the loose gravel. Consequently, I was rather hard on tires, and the purchasing department would use me to test any new brand they were considering. If I could get a year out of a set, the rest of the drivers would probably get at least two. Some tires didn't even make it six months on my truck as I recall, but they must have been cheap imports.

I went through shocks rather fast as well, due to the effect of potholes encountered at high speed. Fortunately I was on good terms with the guys at the maintenance shop, as I'd drop off the occasional stack of skin magazines salvaged from the logging camps for their lunchtime reading pleasure. This meant they didn't make a fuss whenever they were forced to screw, bolt or weld things back onto my truck whenever it went in for repair. One morning when I climbed into my truck after it had been serviced, I found a Do Not Start tag hanging from the steering wheel. When I asked what the problem was, they told me that one of the steering tie-rod ends had come loose and had been hanging on by only one thread. They seemed quite surprised that the truck had actually made it back to town in one piece.

Some problems were a bit more serious, as we discovered when I was travelling down to Fish River with one of the Woodlands staff. For some reason I was driving a pickup assigned to someone else in the company, and I discovered that the vehicle was behaving strangely when going around corners at high speed on the way to the Shelter Bay ferry. The truck kept trying to veer toward the ditch, and my passenger started accusing me of being a lousy driver. I told him that there was something wrong with the truck, but he didn't believe me. We argued all the way to Shelter Bay and continued arguing as we were being loaded onto the ferry. Once parked, I decided to get to the bottom of things and lifted the hood to take a look. I got my passenger to turn the steering wheel as I peered into the engine compartment, and quickly discovered that the steering box was moving as the wheel turned because the truck frame was cracked.

When the welder back in town went to fix the damage, he discovered eleven more cracks in the frame. The person who was the principal driver was told that he'd been driving his truck too hard. He resented this accusation and went around the other trucks in the Woodlands fleet to see what condition they were in. He found cracks in other truck frames, which led to an official inspection of the entire fleet. It turned out that a number of other trucks had one or more cracks in their frames, and the problem was eventually blamed on driving on washboard and other bad road surfaces without weight in the back of the trucks. The bouncing and vibration were causing stress cracks at points in the frames where holes had been drilled.

We found out a bit more about stress and metal fatigue the day my partner and I forgot to take the pickup truck we were driving out of four-wheel-drive after coming out of a gravel road onto blacktop. This resulted in the rear U-joint disintegrating and the driveshaft bouncing along the road as we were doing seventy miles per hour down the Trans-Canada Highway. It was probably a good thing it wasn't the front one, as it might have dug into the road with interesting results. We tied the loose driveshaft up and out of the way with a piece of rope, then carried on to town. Once back in the yard I removed the rope and wedged the driveshaft up underneath the truck, then drove slowly and carefully over to the mechanics' shop. By a stroke of good fortune, the chief mechanic happened to be coming out the door just as I pulled in and applied the brakes, so he heard the driveshaft fall to the ground. After a quick inspection he congratulated me on my amazing luck, since the sudden failure occurred right there outside the shop.

I wasn't alone when it came to problems like flat tires and flat batteries. A logging contractor's crew bus wouldn't start one morning, and we didn't have any jumper cables. I ended up driving my pickup up to the crew bus so that the front bumpers touched, and then we connected the two positive battery terminals together with a steel choker. This actually worked, but the next time I tried to improvise like that, it wasn't so successful. In a benevolent mood one day, I tried to assist a tourist whose vehicle wouldn't start at an Arrow Lakes ferry landing. I cut off two lengths of barbed wire from the top of the fence surrounding a nearby structure, much to the tourist's horror, and attempted to use them to connect my truck's battery to his. Unfortunately it didn't work, so the damage was all for naught.

I eventually purchased a cheap set of jumper cables with my own money and carried them around in the company truck. They were adequate for starting pickup trucks, but not designed for use on logging equipment, as I discovered the day we tried to use them to jump-start a Cat from a skidder. When the Cat operator tried to crank over his engine, the heavy current draw instantly melted the plastic coating on the cables, which fell onto the dirt like water dripping from a washing line. The warehouseman back in town was most sympathetic when I lamented my loss, and he sent me off to the nearby industrial supplier to get a new set made up from heavy-duty welding cable on the company's tab.

•

It wasn't just Cats, tankers and trucks that got into difficulties from time to time. One spring I went into the Glenogle Valley with my supervisor to check snow conditions. We had to run the company snowmobiles on gravel in places where the snow was gone from the logging road, but once we got higher up the snow was continuous. The other machine was up ahead of me, and when I got to a sharp corner in the road, I found its driver standing in the snow with no snowmobile anywhere in sight. He was speechless, and when I asked him where his machine was, all he could do was point over the bank. I looked down, and there it was: hung up in a spruce tree. It seems he'd approached the corner too fast and had bailed off at the last moment when he realized the snowmobile was going to fly off into space. We managed to get it down out of the tree but were unable to get it back up to the road, as the ground was too steep. It would have to be lowered down the hillside, then driven alongside the creek upstream to the next bridge crossing. We didn't have any rope with us, so the two of us rode back to the trucks on the remaining snowmobile and returned the next day to salvage his machine.

We even managed to get a lowbed stuck—on the CPR main line, of all places. This was a serious matter, as the steel body of the lowbed trailer was effectively shorting across the train tracks, which might make it appear on the track monitoring system that a phantom train had suddenly appeared from nowhere. Normally the lowbeds had enough clearance to make it across the tracks without a problem, but this one was a low-slung unit we hadn't used before. Fortunately I had a slight acquaintance with the local CPR Roadmaster, having met him from time to time at local functions, usually near the bar. I called him on the radio telephone to confess what we'd done and promise that we'd be clear of the tracks within a few minutes, while the lowbed driver frantically unloaded the Cat in order to take the weight off the trailer. The Roadmaster generously agreed to turn a blind eye to our situation but warned me that if it happened again there'd be the likelihood of a substantial fine, as track blockages of this kind affect train movements all along the line.

•

I spent a lot of time in my truck, as I'd have a lot of ground to cover in the burning season, what with taking readings from the weather stations and checking slash conditions on the various blocks slated for burning. I'd drive as much as five hundred miles in a day when our operations were widely spread out, and I got so tired sometimes that I'd hallucinate after working a few weeks without a break.

I was driving out of Bush River after many long days when I suddenly spotted a tree lying right across the logging road. I slammed on the brakes and skidded to a stop, only to have it vanish into thin air. A mile or so later I saw another tree on the road and stopped again. By now I'd realized I was seeing things that weren't there, so when I encountered the third tree hallucination, I sped right through it. After a while I started wondering if this was a good idea, as perhaps there might be a real tree on the road somewhere up ahead. The hallucinations got worse, and I started to see some really weird things at the side of the road: stumps turned into strange figures that writhed and twisted as they leered at me. Eventually I ended up pulling over to the side of the road and digging out the filthy sleeping bag that was jammed behind the seat. It was a US Forest Service disposable fireline sleeping bag that I'd scavenged some years before from the IFFS base, and I normally wouldn't have crawled into it in daylight, as it harboured a variety of insect life. That day, however, I was happy to curl up inside it down in the dry ditch next to the truck and get a bit of sleep.

●

Slashburning crewcabs have the following special options, which are not found on regular pickup trucks:

- The floor is shaped like an ashtray and the dash is designed as a boot rest.
- Tire sidewalls are designed to allow for bouncing over logs, boulders and medium-sized roadkill.
- The exterior and interior never need cleaning, and the space below the seat is designed for the accumulation of food waste and empty beer cans.
- They can be driven up to one hundred miles with the low-oil-pressure light flashing.

8

EXPLOSIVES ON THE FIRELINE

My introduction to explosives came when I attended a fireline blasting course in Revelstoke that was put on by the WCB. One of the participants was a forest officer who was recovering from a broken ankle. During the field part of the course outside of town, this caused him some difficulty, as he had to hop along on crutches to where the instructor was demonstrating the correct way of preparing charges. Generally, by the time the poor guy arrived, the fuse was being lit, and he'd have to immediately turn around and hop back to safety before the two minutes was up. By the end of the afternoon he was getting pretty spry on those crutches. The highlight of the exercise was when one of the exploding charges sent a small stump flying directly over the instructor's head.

There was going to be an exam the following morning, and the instructor suggested it would be beneficial to study the written material carefully that evening. The other two attendees from Evans and I went straight to

Top: A Sky Spider descending from a hovering helicopter during rappel practice at the Golden Airport.

Bottom: Miscellaneous debris that accumulated at the Tsar Creek fly camp after several days of timber cruising.

Water being delivered to the pump setting above the Double Eddy Creek fire, at great expense, by a helicopter equipped with a monsoon bucket.

Top: Ignition has started on this block at Glenogle Creek. The crew is lighting parallel strips with hand-held driptorches.

Bottom: Light-up is progressing, and the lines of fire are joining up. The hot fire at right is starting to create indraft.

Top: Light-up is complete. The crew is monitoring the burn from the safety of the catguard. Note how the vegetation in the foreground is bending from indraft wind.

Bottom: View of a burn underway on a block close to Glacier National Park. The main fire is drawing the fire at right in and away from the catguard along the timber edge.

A convection burn is taking place on this block in the Beaverfoot Valley. The helitorch is laying a line of napalm as the helicopter flies around the central fire.

A helitorch lighting slash in late afternoon in the Bush River valley. The block is in shadow, while the sun is still shining on the mountain ridge.

Top: A spring burn on a Copper Creek high-lead block. The helicopter (barely visible as a white dot at lower mid-left) is laying down another line of fire.

Bottom: Light-up is complete. Heat from the fire could be felt at the camera location.

An escape that occurred during a slashburn on a steep high-lead block in the Glenogle Valley. A slight "overachievement."

This fire devil appeared after dark during a broadcast burn. The upper part is just visible at top of photo. Burning debris is whirling around the central flame column.

Top: A homemade flamethrower being tested after dark. Note observer at lower right.

Bottom: Another drum of napalm is being mixed while the helicopter waits with blades turning. Safety precautions are noticeably absent.

Top: A Wajax pump in operation, pumping water from a makeshift pond that contains water of dubious quality.

Bottom: A Ukrainian Water Bomber heading up a catguard on the Double Eddy fire with yet another thousand-gallon load of water.

Top: A D7 Cat stuck fast at Tsar Creek on a logging road that proved to be a bit soft.

Bottom: After being pulled out of the mud, the D7 ended up in Kinbasket Lake a few days later. The diver is about to release the brakes.

Top: Field testing a landing pile ignition device consisting of a five-gallon pail of napalm wrapped with detonating cord.

Bottom: A six-wheel-drive tanker being assisted across Bush River at the end of spring burn mop-up operations.

Top: A landing on a high-lead block being opened up with dynamite in order to expose residual fire burning beneath the dirt.

Bottom: This is what remained after we blew up an abandoned school bus. One sizeable fragment is visible at the timber's edge.

Top: While waiting for me to arrive by helicopter, the burning crew thoughtfully marked the landing spot so it would be easy to see from the air.

Bottom: A landing pile at Smith Creek that has been lit up and is radiating enough heat to ignite several nearby snags.

Top: Looking into the heart of a burning landing.

Bottom: A spot fire has started on the right-hand side of the road. The crew is going to see if there is more fire across the catguard.

the bar of the Sandman Inn, where we were staying, for a group study session and beer. The general manager of our company happened to be in town also, and he stopped by that evening and bought us a drink. I'm not sure what he thought about the prospect of the three of us being entrusted with explosives. The instructor passed by us later on the way for a swim, as he was staying in the same hotel, and he asked how the studying was going as he eyed the collection of full and empty glasses on our table. Our reply was that we were confident we had everything figured out. We were still there when he got out of the pool and went for supper in the restaurant. By the time he'd finished his meal, we still hadn't moved, apart from the odd trip to the washroom and occasional waves to the barmaid. I seem to recall that he was shaking his head as he went past our table, which was now awash with stale beer and studded with soggy cigarette butts. We all managed to pass the test the next morning, however, which was miraculous when you consider that we'd stayed in the bar until they threw us out at closing time and never did get around to reviewing our notes.

The class of blasting licence I received as a result of this course restricted me to using detonating cord only, the idea being that it would be used for blasting fireline and taking down dangerous snags. It wasn't long before I decided that I wanted to get a better class of licence that would permit me to use dynamite, so I eventually wrote another exam to obtain it.

I started using explosives to blast open landing hangover fires, as it made the task of extinguishing them go a lot faster. Prior to this, the method was to open them up with water, rather like a hydraulic sluicing operation. In most cases there wasn't a water source close by to run a pump from, so water would have to be hauled in by tanker truck. Since many landings were up on mountainsides, this could be a slow process. I discovered that a few sticks of dynamite in the right place could save a lot of tanker trips by removing the dirt overburden to get at the burning material underneath. The only drawback was that sometimes the fire burning underground was hot enough to raise a concern that the dynamite might go off prematurely as it was being loaded. On consulting the CIL's *Blasters' Handbook*, I found reference to an explosive called Pyromex, which was designed for blasting operations at steel mills, as it was tolerant of high temperatures. When I phoned CIL to see about buying some, I was told that they no longer made the product. They asked what I wanted it for, and when I explained to them what I was doing and why I needed a heat-resistant explosive, they told me that what I was doing was "extremely inadvisable."

Eventually I came up with a solution: I put the dynamite in plastic bags full of water.

I never did have any dynamite go off prematurely due to heat, but I have had a few anxious moments. In one instance a charge failed to go off when I was blasting a landing hangover, as somehow the detonating cord had burned through. I ended up retrieving the dynamite with a shovel and watching as it sizzled like a fried egg on the shovel blade.

I have used dynamite to cook with, however, and it does an excellent job. I'd set fire to a piece of it on the ground, surrounded by three rocks to support whatever was being cooked (usually either canned stew or Kraft Dinner). Dynamite burning in this manner generally won't explode, but the practice is inadvisable, as an accident could spoil the cooking.

•

An interesting moment came during mop-up operations on a spring broadcast burn when I had to bring down a burning spruce snag that was too dangerous to fall with a chainsaw. One of the crew was with me as I sized it up, and we could see that it was hollow and blazing merrily inside. Before I attached the Primaflex detonating cord, I thought it would be prudent to extinguish as much of the fire inside the snag as possible. All we had with us was a piss can, which wasn't very effective. We squirted water as far up as we could reach, and it seemed that the worst of the fire was out, so I wrapped a few turns of cord around the snag and took a cap and fuse assembly out of the cap box. By the time I turned back to the snag, the fire had flared up inside again. My partner pointed to the Primaflex and exclaimed: "Look at that!" The flames had melted the plastic outer coating, which was dripping off to expose the fabric underneath. Not a healthy sight, as once the fabric burned through, the flame would be in direct contact with the explosive. I told my partner to get the hell out of there, taped the cap to the end of the Primaflex rather hurriedly, then held the end of the safety fuse inside the snag just long enough to be sure it was burning. I then joined him some distance away, and together we watched the blast take the snag down in a shower of flaming shattered wood. I didn't have a watch, so I had no way of timing the fuse to see if it was actually two minutes before the explosives went off. I was, and still am, curious to know whether the detonating cord exploded prematurely due to being set on fire, and whether I should have struck off another of my nine lives that day.

Sometime after that I actually had the end of a roll of detonating cord exposed to fire. I'd left it on the ground along with sundry other blasting material while I did a reconnaissance of the edge of a slashburn escape. By the time I returned, the fire had crept much further than I'd expected and had reached my temporary explosives cache. The loose end of the roll of E cord was now in the flames. I used this type of detonating cord for connecting charges together, and it's a lot less powerful than Primaflex. Still, it was a one-thousand-foot roll and would have made a healthy bang had it gone off. I watched with interest for a moment to see if it would actually start burning, then decided I'd better cut off the scorched end. Stamping on detonating cord isn't a good idea, as under certain conditions it may explode. It's perfectly safe to handle otherwise, and in fact I once used a piece of it to hold my pants up after my belt broke as I was doing some blasting work on a fire. It wasn't entirely satisfactory, however—some of the explosive powder within it trickled out the ends and got into the teeth of my zipper.

I didn't have any need to use detonating cord to construct fireguards on any of our slashburn escapes, but I did get asked to blast some for the Forest Service on one of their wildfires. They arranged for me to drive out and meet up with a helicopter that would fly me and the explosives to the site. Once I arrived at the helipad, I was about to load the reels of Firecord into the cargo compartment when the pilot stopped me. He wanted it all loaded into a cargo net, along with the cap box, as he'd prefer to sling the stuff into the site rather than carry any of it inside his helicopter. I explained that it was strictly forbidden to carry the blasting caps in with the Firecord, as that could lead to a nasty accident. His reply was that if anything went wrong, he'd just let the sling drop. I pointed out to him that his reaction time couldn't possibly be fast enough, given that the detonating speed of the explosive within the blasting cord is in excess of twenty-two-thousand feet per second. Firecord is a type of detonating cord made specifically for blasting fireline. It has a layer of fire-retardant chemical inside that prevents the explosive from starting fires when it's detonated near dry fuels. With twice the explosive power of Primaflex, it's roughly equivalent to one stick of dynamite per foot, so the load we'd be carrying would have vaporized the helicopter if it were to go off accidentally.

We finally reached a compromise: he flew the sling load of Firecord to the site, then returned for me and the cap box. Even though I'd assured him the blasting caps were the safety fuse type and couldn't possibly be set off by static electricity or the radio transmitter in his helicopter, he seemed very nervous as we flew up the mountainside. I never saw him again—for some reason he arranged for another machine to pick me up when the job was finished.

We packed the Firecord down to where a crew was hard at work digging a fireguard along the side of the mountain. The plan was for me to speed up the process by blasting the line where the ground was particularly difficult to work on, so I unrolled the cord with the help of the crew. In some places, there were logs lying on the ground that needed to be cut so that the cord would lie along the ground for a more effective result, and there were a couple of saw operators to handle it. One of them was sitting on the ground and filing his saw chain when I came up to him, and he asked me if he was in the way. I replied that he was, but that the cord would take care of it, and I wrapped it a couple of times around his outstretched leg. He didn't see the joke for some reason. I believe we eventually ended up with nearly a thousand feet of Firecord stretched out along the mountainside. The bang it made when set off was probably heard over in Revelstoke.

•

I was flown in to blast a burning snag on another Forest Service fire up the Sullivan River. The snag was apparently too dangerous for the fallers to deal with, and when I looked at it, I decided that even I wasn't crazy enough to attempt the job, as it was a large hollow snag with fire up inside it and burning chunks falling off at regular intervals. I told the fire boss that it would come down on its own in a couple of hours, and that he should just cordon off the area in the meantime and keep everyone well away. The flight crew and I found the time to accept a meal at the fire camp for our trouble, and then we took off again. The pilot of the helicopter was very disappointed that I didn't blast the snag—he'd been really looking forward to seeing something blown up. As we were flying over the lake at the time, I offered to drop a charge out the door, and as we hovered at least five hundred feet above the lake, I primed a stick of dynamite. The flight engineer was leaning out the open door expectantly as I lit the fuse and tossed out the charge. He didn't bother to secure himself with a safety line, which made me a bit apprehensive when the pilot banked the helicopter sharply so that he could get a good view of the bang.

Unfortunately the dynamite fell so rapidly that the cap and fuse were ripped out by the wind, so there was no visible explosion. Probably a good thing, as there were monitors along the lake to detect any landslides that might fall in and generate a tidal wave, which might cause problems at the dam holding the lake back. This was discovered accidentally by a log salvager who decided to do a bit of fishing one day with a "Public Works minnow" (dynamite) as he was travelling up the lake in his boat. I don't think any fish floated to the surface, but very shortly afterwards a helicopter appeared over the hill from the dam to look for the landslide, as their sensors had registered his fishing attempt. It didn't take a Sherlock Holmes to connect the salvager with the still-visible water disturbance in the boat's wake.

After our attempt to make an underwater explosion failed, we flew over to the Selkirk Mountains and dropped a couple of charges into crevasses at the head of a snowfield. They went off this time, as we were hovering much lower when I dropped them out, but we were unable to start any avalanches—it was late summer, and the snowfield was stable. After this, the pilot showed me how he could write his first name in the snow with the helicopter skids. As he manipulated the helicopter in dizzying circles to demonstrate his penmanship, I was thankful he had a short name, as I'd neglected to take a Gravol pill before the flight.

•

I went over to the company's operations in the Fish River valley with one of the logging supervisors in order to blast a snag that was too dangerous to be felled with a saw. The trip involved a ferry crossing, but unfortunately there wasn't a dangerous cargo sailing that day. We didn't want to wait until later in the week for one, so I covered the case of dynamite with my jacket and put my feet on it while we were loaded onto the ferry. My partner was a bit apprehensive about the consequences of being caught, but I wasn't worried, as I'd previously hauled gasoline across on the same ferry illicitly when burning in Fish River. Whenever we were asked what was in the tank, we'd tell them it was diesel, and hope to hell that they wouldn't test the contents with a match.

Once we arrived at the snag, my partner asked me how much dynamite I'd be using. I told him all of it, as that way we wouldn't be forced to violate safety regulations a second time on the return trip. The resulting blast removed all traces of the snag—not even the stump was left, and the rock face behind the snag had been shattered. Presumably the pieces landed somewhere further up the valley.

I took down quite a number of burned-out snags and hollow cedar trees with explosives, particularly up Bush River. It was much safer than cutting them down with a saw, and I often used a combination of Primaflex detonating cord and dynamite. I'd put a few wraps of Primaflex where I wanted to cut the snag or tree, then attach half a stick of dynamite further up in order to give a bit of a push. The two charges would be connected with E cord, so they'd go off together. Now and then I'd shinny up the tree or snag in order to place the dynamite, and as this required both hands, I'd tie the end of the E cord to my belt so it would be readily available.

On some escapes there'd be so many snags in an area the crew was about to go into with hoses that I'd connect a number of them together as I set the charges, so they'd go down like dominoes when the blast went off. The air would be full of flying splinters, and the shock wave would test the stability of all the burned snags outside the blast area.

Sometimes the crew came with me when there were snags to be blasted, so they could watch the fun. On one block we had to climb above the top edge of the burn to take down a few burning snags where there'd been some fringe damage. There was still snow on the ground under the trees—it was a high-elevation spring burn, and the fire had burned through the crowns overhead when it escaped the block boundary. One of the crew was packing the roll of Firecord I was going to use for the job, and when we stopped for a short rest, he dumped it onto the snow and then asked me how safe the stuff really was, as he was a bit uncomfortable carrying high explosives. I told him it was perfectly safe unless it got cold and wet, then turned to look at him with a look of simulated horror, exclaiming, "Oh shit, you've dropped it in the snow!" as I backed away nervously. We all had a good laugh watching as the poor guy frantically brushed off the snow sticking to the roll, with a look of panic in his eyes. He refused to carry it after that for some reason.

One of the snags I blasted that morning was quite spectacular: after the cord detonated, the upper part stayed in the air for a second before crashing to the ground. It would have made a great photograph if I'd had a camera, and I received a round of applause from the crew.

•

Opening up underground fires could be interesting at times. I went down to a timber sale near Invermere with one of the logging supervisors to deal with a couple of landing hangover fires. We took a slip-on tank unit with us, as well as some explosives to open the landings up so the water would go further. I set the first charge, then drove away to a spot I knew would be safe. My partner disagreed, telling me that we should be further away from the landing. I tried to explain that any debris thrown out by the blast would follow a curved trajectory, and that we were parked more or less directly below the zenith of this trajectory. There was a heated argument while the fuse was burning, until I suddenly said, "Have it your way, then!" and drove to the spot he insisted would be safer. The blast went off, and we watched a small black speck appear in the distance, headed straight for us. The speck became larger very rapidly, and I watched with much satisfaction as it turned out to be a sizeable chunk of charred wood that hurtled to the ground uncomfortably close to the truck we were sitting in. For some reason, my partner's face appeared to be a rather odd colour as he stared at the impact site.

One of the burning crew was with me when I was blasting open a landing hangover fire up Bush River. As usual, neither of us was wearing a hard hat, so we took shelter behind a small rise in the ground after the fuse was lit. There was the usual loud bang, followed by a whizzing sound as a rock the size of a grapefruit embedded itself in the ground between the two of us. My companion looked at the steam rising from the hole and then turned to me slowly as he commented, "A fellow could get a headache from one of those."

A hard hat probably wouldn't offer adequate protection to fly rock of that size descending vertically, but sometimes someone on the crew would actually wear one when close to a landing blast. One time that springs to mind is when two of us were standing behind the Kenworth tanker, waiting for a charge to go off. My partner had dug a filthy hard hat out from the debris in the tanker cab and put it on for protection. Personally I would have been more worried about contracting scabies from that loathsome piece of headgear. I never bothered wearing a hard hat when I was working with explosives, going on my usual assumption that the Devil looks after his own. As the hat wearer was looking toward the landing, prepared to dive under the tanker if anything really big headed our way, I picked up a chunk of rock, and a fraction of a second after the blast went off, I hit his hard hat with it. The result was most gratifying: he sank to the ground, speechless with shock. His first words once he got his voice back, as I recall, were, "I think I shit myself."

Probably the worst choice of protective headgear I witnessed was when a landing hangover I blasted open produced an unusual amount of debris heading skyward. One of the burning crew was with me, and as we waited for it all to return to earth, he held the cap box he was carrying over his head protectively. It would have been interesting if a rock had landed directly on the box, as there were at least twenty blasting caps inside it.

Using explosives to remove snow was a little safer: whatever was blown up into the air came down as a fine shower of snow and water droplets. One spring I needed to open up access to the back end of the Blaeberry Valley, as I was planning to do some spring burning in that part of the country. The company didn't want to spend money on a Cat or grader for clearing the snow slides that were blocking the road, so I went out with dynamite to get rid of them. I buried sticks at regular intervals across the compacted snow, the idea being that the craters produced would allow the snow to melt much faster. I set all the charges, connected them together with detonating cord and was just about to light the fuse when a pickup truck drove up. I waved it back urgently, then lit the fuse, and after a couple of minutes there was a nice loud bang that echoed around the valley. When I went to see who had driven up, I discovered they were hunters who were hoping to find grizzlies further up. As the echoes died away, they complained that I'd probably scared them all out of the valley, and then they turned around and drove away. I was really happy that I'd ruined their day—I detest trophy hunting.

Now and then we'd get a lightning storm moving through while we were engaged in firefighting activities. Normally that's not a problem as long as you're careful not to be in a patch of snags or unstable timber, as there can be sudden violent downdrafts accompanying a storm that seem to blow from all directions except straight up out of the ground. One lightning storm came through while we were mopping up a slashburn near Moose Creek in the Beaverfoot Valley. This area seemed to get a lot of electrical storms for some reason, and I'd witnessed some fairly intense ones there in the past. This time, however, I had a steel magazine in the back of my pickup that contained six cases of dynamite, which is not the kind of cargo you really want to have behind you when an electrical storm passes overhead. As I sat in the truck, looking down at the cutblock we were working on, I saw a lightning bolt strike a tree, below my elevation and not that far away. At that moment I decided to move further downhill, and I drove to where the crewcab was parked while the crew was eating lunch. I pulled in behind them as more lightning hit nearby, with accompanying claps of thunder. They called me on the radio, asking me to please go away, as they preferred not to go up with me and my truck if lightning was to score a direct hit on the magazine. When I refused to move, they started up and drove off, but that didn't help them, since I was following right behind as I accused them over the radio of being overly paranoid.

This wasn't the only time I was near an electrical storm while carrying explosives. The next occasion happened when I'd been blasting snags near the top of a burn escape in the Glenogle Valley. I was starting back down the mountainside with a partial case of dynamite under one arm, the cap box and a fire shovel in my other hand,

and a radio on my belt, when I realized that the air was so electrically charged that the hair on the back of my neck was beginning to stand up. It's surprising how fast you can move down a mountainside when you've got an incentive.

I nearly had a nasty accident on the same mountainside the following day. Once again I was coming downhill after blasting more snags, only this time I slipped on a moss-covered rock, fell over, and started rolling downhill. I let go of the cap box, dynamite case and shovel (in that order) so I could concentrate on not bashing my head against anything hard. As I tumbled, I noticed something shiny appear at my throat each time I rolled. When I finally managed to stop myself, I discovered that the shiny thing was a blasting cap. I'd inadvertently left a cap and fuse assembly around my neck. Normally they'd be kept in the cap box, but that day I was wearing caulk boots, and I had been hanging the caps around my neck while I was setting charges rather than setting them on the ground, concerned about what might happen if I accidentally stepped on one with my spiked footwear.

I wasn't the only one to make a small mistake with blasting caps. A member of the burning crew was working toward getting a blasting licence, and I was letting him set charges from time to time to help him gain the necessary experience. On one of these occasions I caught him just about to hold his lighter to the wrong end of the cap and fuse assembly that he'd inserted into a stick of dynamite. One end has the silver-coloured blasting cap and the other end has a copper-coloured Thermalite igniter that has about the same dimensions. To light the fuse, you pry open the igniter and hold a match or lighter to it. In order to give you a hint as to which end is which, the manufacturer thoughtfully marks the cap with "Danger: High Explosive" or words to that effect. When I pointed out that what he was about to do would probably blow a few fingers off at the very least, the colour drained out of his face and both his hands shook a bit.

He did eventually obtain his blasting licence, though I don't know if he picked up any bad habits from watching me handling explosives. I'm not sure if the WCB would have been overly thrilled at the thought of me teaching others about blasting, as I couldn't resist the urge to attach explosives to anything flammable, such as napalm containers, propane bottles and spray cans containing paint or ether. (I was unable to persuade the millwrights at the plywood plant to let me have one of their hydrogen cylinders, as they knew exactly what I planned to do with it.)

•

Not all the blasting I did was directly related to fire control. There was an old school bus up Bush River that had been used by a logging contractor for accommodation and subsequently abandoned. At some stage someone had helped themselves to the wheels and axles, so it was now a complete write-off. We'd often driven by the remains while working in the area, until one day somebody on the crew suggested it might be fun to blow it up. I can't remember how many sticks of dynamite we shoved underneath it, but I saw evidence of porcupines living there, so I added a couple of extra ones, just in case they were still around. I went up the nearby spur road to block it off in case anyone might be coming down it as the fuse was burning. The blast blew the bus apart, sending pieces high in the air. Some of the interior woodwork ended up hanging in a spruce tree, along with most of the insulation from the walls, making it look a bit like an environmentally unfriendly Christmas tree. A large piece of the bus landed in an adjacent cutblock that had recently been planted, which led to a complaint from the silviculture survey crew when they discovered that it was sitting on top of some little seedlings they were supposed to count. For some reason I was automatically blamed for the explosion, and I was forced to divert a bulldozer that was heading up the valley to remove and dispose of the wreckage.

The company road construction superintendent asked me to blow up a beaver dam near a logging road that had been causing a problem (the dam, that is—logging roads never cause problems). The resident beaver had been constantly plugging up culverts under the road, which was causing washouts. The company had tried other methods of preventing the road damage, such as fixing gratings over the ends of the culverts, but nothing had worked. Finally the Ministry of Environment reluctantly agreed to let the company blow up the dam, but on the condition that the beaver lodge not be disturbed.

I was most enthusiastic when I got out to the site, as I'd always wanted to blow up a dam after watching the movie *The Dam Busters* on television. Admittedly this dam wasn't exactly a German hydroelectric structure, but one has to start somewhere. I ended up attaching numerous sticks of dynamite to the end of a long pole so that I could place the charge at the bottom of the pond next to the base of the dam. As usual, I wanted to get a good view of the blast, so I didn't back off as far as I should have. When the blast went off, there was a column of water, mud and debris heading skyward that was definitely impressive. Unfortunately it all had to come back down somewhere, and I suddenly realized that a sizeable chunk of wood was headed straight for me. I managed not to be in the same spot it picked to bury itself in the ground, and I went to look at the results. The dam was no longer structurally sound, and the water from the pond was tearing away those parts that the dynamite hadn't managed to dispose of. It was suggested to me that it might be wise to blow the lodge up as well while we were there. I was tempted to ask about the MOE stipulation, but I decided that it wasn't advisable to question a superior, and besides, orders are orders. Accordingly I tied a couple of sticks of explosive to a chunk of wood, lit the fuse, and then steered it into the beaver's residence. This structure disappeared skyward in a much smaller column of water, mud and sticks, but I doubt the beaver went up with it. He'd probably packed his bags and cleared out after the first blast.

While the local golf course was under construction, I was asked to blast some stumps in a swampy area where the excavator was having difficulty operating. It wasn't a bad job, apart from having to put up with the swarms of mosquitoes inhabiting that part of the swamp. I loaded explosive charges under a number of stumps and connected them all together with E cord, then attached a cap to the end of the line, lit the fuse and walked away. As I was leaving, I reached down automatically to check that I had my blasting pliers and found that the pouch was empty. I knew I'd probably left them on one of the stumps, but as I wasn't sure which one, I had to run around and check them all. At one point I realized the fuse was burning ever closer to the blasting cap, so I really needed to find where that was before the whole lot went up and took me with it. I found the cap, yanked it free from the detonating cord and threw it away. Naturally the pliers were discovered sitting on top of the last stump I checked.

●

During a mop-up operation well up a mountainside, I played a mean trick on some of the crew. There was a small creek on one edge of the burn that didn't quite have enough water in it to run a fire pump, so I'd been working my way uphill, blasting holes in the creek bed. These would fill with water and give an adequate supply to run the small Shindaiwa pump. The crew was working their way up the creek at a safe distance behind, setting the pump in each pool and spraying water until they reached the limit of the pump's efficiency to push water uphill, at which point they'd move to the next pool upstream. Once I'd completed the last blast at the top of the burn, I headed back downhill and crept up to where the crew was working. As I hid behind a tree, I attached a piece of fuse to a stick of dynamite (minus the cap, of course), lit the end, then threw it over to the crew. It landed on the ground right beside them, and they turned to see what it was. The instant they noticed the smoke coming from the burning fuse, they dropped everything and bolted to safety. Their faces peered out anxiously from behind some trees as the fuse burned down to the dynamite. It didn't explode, of course, and eventually they calmed down and saw the funny side.

    A day or two later I was sitting in my truck with one of the crew after we'd been doing some mop-up work that had involved the use of explosives to take down a few burning snags. For some reason there was a cap and fuse assembly hanging from the heater controls on the dashboard, and my passenger eyed it with interest as he asked how powerful blasting caps were when they went off. I suggested that he might like to find out for himself, and I promptly held a lighter to the end of the fuse. We both sat there in silence as the fuse burned slowly but surely toward the blasting cap. Finally he asked me if I was going to put it out, and I replied that I wouldn't dream of it, as I wanted to let him see what would happen. At that he shouted, "Holy shit!" and bailed out the door, landing in the ditch, which fortunately was empty except for all the litter he'd tossed out the window during the lunch break. While he was picking himself up and swearing at me, I threw the fuse out my window before the cap exploded.

I got an equally gratifying reaction the day I drove up beside the crewcab when it was parked on a landing as the crew was eating lunch. While I was sitting in my truck, talking across to them through the windows, I cut the blasting cap off a cap-fuse assembly, taped the fuse to a large package of matches, lit it, then threw the whole works into the crewcab. They looked at the smoking fuse for a second, then suddenly all the doors flew open and the crew bailed out onto the dirt, accompanied by an assortment of bottles, half-eaten sandwiches and pornographic magazines.

•

Once or twice I was given old explosives for disposal. Some of it was pretty dubious-looking stuff, particularly if it had been frozen for a couple of winters, but I never refused to take any of it. After all, someone had to get rid of it, and there's nothing like going out into the woods and making a nice loud bang. One old case of dynamite I received looked like it wouldn't stand up to too many miles of bouncing about in the back of the truck, so I only drove as far as the gravel pit just beyond the start of the Beaverfoot logging road. The bottom of the pit was full of water, so I placed the dynamite on an old wooden pallet, then pushed it out once the fuse was lit. The explosion emptied most of the water out of the pit, though a lot returned as rain mixed with small fragments of cardboard and pieces of kindling.

I took another bundle of old explosives out to the West Columbia wrapped in my jacket, as the dynamite had been frozen over the winter, which might have made it over-sensitive, and the jacket kept it from rolling about on the floor beside me whenever I hit a pothole. Just after I'd driven across the bridge at Donald, west of Golden, I was flagged down by an RCMP officer. I was apprehensive at first—along with the explosives on the floor was a bottle of rum I was taking along for emergencies, and I hoped he wasn't about to do a vehicle check. As it turned out, all he wanted to know was if I had any tools that would help open the doors of a car that had crashed into the ditch just around the corner. There were a couple of injured passengers they couldn't get out, as the doors had been damaged in the crash and wouldn't open. I could have offered to blow the doors off but decided they probably wouldn't take me up on it. We ended up trying to saw through one of the door catches with a blunt hacksaw blade that somebody produced, but progress was painfully slow. The ambulance finally turned up with the Jaws of Life, but they weren't able to get it working, so we carried on sawing. We finally cut through, and several of us managed to bend the damaged door back far enough to get the injured occupants out of the vehicle. Once they'd been loaded up and hauled away, I carried on to the West Columbia and found a quiet spot to detonate the explosives. When I talked to one of the paramedics back in town the next day, I was told the reason the Jaws didn't work was that they'd hooked up the hydraulic connections incorrectly.

Not everyone wanted to get rid of explosives. While we were conducting slashburn mop-up operations in the Bush River valley, one of the crew came up to me one day and asked if he could have some dynamite. I refused, naturally, as he had no business with the stuff, and inquired what he wanted it for. He wouldn't tell me at first, but eventually I discovered that he was up on an impaired driving charge (not his first) and had been advised by his lawyer that if he could obtain some explosives, he might be able to work a plea bargain with the RCMP. At that time, such deals were possible: lead the cops to a handgun or some explosives and they just might tear up the charge sheet. This didn't do much for my confidence in the ethics and probity of lawyers, and from then on, as a precaution, I kept the key to the powder magazine in the back of my truck under my pillow at night while I was staying in the Bush River logging camp.

It was in this camp at about the same time that I was sitting in my room doing some paperwork after supper while some of the crew were drinking in the next room. I was trying to tune out the loud voices and obscenities echoing along the corridor, but I decided I had to draw the line when they began insulting my New Zealand heritage, accusing me of close encounters with wool-bearing animals, among other things. Rather than go in there and deny the slanderous remarks, I went out to my truck and took a stick of dynamite out of the magazine. I cut the blasting cap off a cap and fuse assembly, stuck one end of the fuse into the dynamite, lit it, then tossed it into their room, saying, "Here's a present from New Zealand!" The ensuing scene was like something from a Saturday afternoon cartoon show: they all tried to get out at once, and two of them were actually jammed in the doorway at the same time, still running. When they finally returned to a room that stank of burned black powder, one of them said that he didn't think I'd actually go so far as to blow up the bunkhouse, but he didn't want to stay there to find out.

Our main explosives magazine would be inspected from time to time by the RCMP to check that everything was in order, and that the quantity of explosives in the magazine tallied with the totals shown in the inventory book. They very rarely did, due to sloppy record keeping, as I would have to explain each time. During one of these inspections I was moving a case of explosives as instructed by the attending officer when we discovered it had leaked some viscous fluid onto the magazine floor. I knew the stuff was quite harmless, as that particular explosive didn't have nitroglycerine as an ingredient. The officer didn't know that and was visibly concerned when he saw what was on the floor. He asked what it was, and I replied that I wasn't exactly sure, but I hoped that a fly didn't land on the stuff and set it off while we were there. At that point he suddenly cut short his inspection, and on his way out the door he told me to get it cleaned up immediately. I promised I would and added that if he heard a loud bang as he was on his way back to the police station, he'd know exactly what it was.

●

Interesting fact:
The Finnish Army term for detonating cord (Primacord) is *anopin pyykkinaru*, which translates to "mother-in-law's clothesline." (It does look a bit like clothesline.)

9
MISCOMMUNICATIONS

When I was working as an initial-attack crewman in 1977, we'd often be on standby in the evenings and on weekends and be expected to respond to a fire on a moment's notice. We were issued portable radios so that we could be contacted if we were in town somewhere when a fire was reported. These were Motorola PT300 models, which were referred to as lunch-pail radios, as they did look very much like lunch pails and were fairly heavy—they contained eleven D cell batteries. They were also quite bulky, and whenever several of us were sitting in the local pub waiting for lightning to hit a tree and provide us with some overtime, there'd be less room on the table for beer glasses. Not that we could afford many of those, as standby rate was a third of our meagre hourly rate, and we'd calculated that we were sometimes drinking our pay faster than it was accumulating.

Whenever we were fighting fire, we found the PT300 to be more of a nuisance than an asset, as you needed one hand to carry it, and we were packing enough gear as it was. When you're dragging around a few hundred feet of wet firehose or hauling a fire pump and fuel container up a mountainside, you don't feel like lugging a radio along as well. This meant they would often be left sitting on a stump or next to a fireguard, and as the colour of the radio blended in nicely with the surrounding vegetation, they'd sometimes get mislaid. It wasn't unknown for one to get destroyed when a fire took off suddenly or a tree was accidentally felled directly on top of it.

One of our firefighting crew liked to pack a portable radio around town because it made him look important. He went off to have lunch at a restaurant one day, complete with a lunch-pail radio. I knew exactly what his routine would be when he sat down at a table: he'd turn the squelch control until it made a noise that would attract the other diners' attention. Presumably this would give him standing and respect as a Man on Call. I took my portable radio into the bathroom of the motel unit we were staying in and, when I estimated that he would have attracted everyone's attention, pressed the microphone button as I held it in the toilet bowl and flushed. It worked—when he returned after lunch, he called me an asshole, as the volume had been turned up on his radio and the restaurant had echoed with the sound of a toilet in operation.

Eventually the surviving lunch-pail radios were phased out and replaced by smaller MH70 units that could be carried on your belt. The ones we were using still only had two channels, but they made life much more convenient, as we could have them with us at all times when working on a fire. They also took up less room on the table in the pub and were very waterproof, as we discovered whenever beer was accidentally spilled on one. They were resistant to other liquids too, as I discovered the day I wasn't able to get out of the way in time when an air tanker was dropping a load of retardant on a fire. Both my radio and I received a coating of the thick, slippery red liquid, which doesn't taste very good. The radio worked fine once I'd scraped off the worst of the mess and found the transmit button.

They also proved to be very tough radios the day I was riding as a passenger on a bulldozer. The ground we were travelling over was extremely rough, and we were being bounced around quite a bit, so much so that my radio fell out of its pouch and landed on the tracks. By the time I could get the operator to stop, we'd already run over it and I expected the worst. To my surprise, when I dug it out from where it was embedded in the ground, I found that it was undamaged and still worked perfectly.

There were limits to the survivability of portable radios though, and I found out during a slashburning operation that the later model MX350 wasn't fireproof. I'd been clambering through heavy logging slash with a driptorch, lighting as I went, when I fell over and the radio came out of its pouch (the retaining strap was broken). I tried to find it amongst the branches and other debris, but the fires I'd lit were starting to spread and my escape route was about to be cut off, so I didn't have much time to search. If I'd stayed a bit longer, I might have found the radio, but I wouldn't have made it out in time. It was a tough decision, as the MX350 was a very expensive radio back then. To this day I feel bad about it burning up, as that was the only thing I ever lost during the years I spent working in the woods, unless you count all the blood siphoned off by mosquitoes and blackflies.

I went back the next day and discovered the charred remains of the radio, which I carefully collected and placed in a box to take back to the company. This way they'd know that the radio really had been destroyed and wasn't just sitting in a pawnshop somewhere. I handed the box of fragments to the purchasing manager and asked him if he'd mind sending my radio to the repair shop, as it didn't seem to be receiving very well. I can't remember what his exact reply was, but I doubt if it was printable.

My handheld radio went everywhere with me when I was on standby—it went to all the best parties and a lot of the worst ones as well. Now and then I'd find that I was unable to leave a party when it was over due to lack of motor coordination, and I would have to search the place for my radio once I'd finally woken up in an unfamiliar environment. I came up with a routine to make it easier: when I arrived at a party, I'd leave my radio and truck keys inside one of my boots near the front door. That way all I had to do was find my way to the door in the morning and I'd be all set to go.

A radio is a vital tool on a forest fire, particularly if you've been dropped into a remote location. You can call in more resources if needed, and you can get an injured firefighter flown out quickly if an accident happens. It's also a way of getting out an urgent message, as we were forced to do on a fire we were fighting up Bush River.

Three fires had been started by a passing lightning storm, all of them on the mountainside with no road access. The only way to attack them on the ground was by taking a helicopter in, so two of us were flown to one of the fires, along with our equipment. The handheld radio we had with us had only company channels, which gave us communication with the office through the company's mountaintop repeater. Soon after we arrived at the fire, we heard the unmistakable sound of an air tanker flying up the valley. We assumed it was going to drop retardant on one of the other two fires nearby that were as yet unmanned, but we started to have doubts once we realized it was heading straight for us. There had obviously been a mix-up in the Forest Service fire control room, and we had no way of communicating with the bird-dog plane accompanying the air tanker, as our radio didn't have any Forest Service frequencies.

The fire we were working on was in heavy timber, so it was unlikely the bird-dog officer would see us as he flew over, and we had no signal flares to attract his attention. I called the company office on the radio to get them to relay a message through the Forest Service dispatcher, to tell the planes that they were heading to the wrong fire. In the meantime my partner frantically gathered up our food and personal equipment to hide it under a log so that it wouldn't be covered in a layer of retardant if they did drop a load on us. Luckily the message got through at the last moment and the planes veered off.

That was when I swore I'd never get caught like that again and became interested in radio frequencies, which led me to purchase a scanner so I could amuse myself by figuring out who was using which radio channels in and around town. It became quite a hobby after a while, and I came up with a number of different strategies for obtaining the information I wanted. Eventually I bought a programmable handheld radio so that I'd have the ability to call out on virtually any channel in the event of a dire emergency in the woods. A couple of the crew also became interested in this type of radio, which wasn't necessarily a good thing, as it turned out.

I was a passenger in the crewcab heading off somewhere south of town when we were stopped by the flag girl of a road repaving crew. To amuse myself while we were waiting, I took out my portable frequency counter and found out what radio frequency was being used by the paving crew. Once I found it, I made the mistake of letting the crew see what it was. One of them immediately programmed it into his handheld radio and started chatting with the flag girl. She had no idea who was talking to her, as it obviously wasn't someone on her crew, but she didn't seem at all upset. Unfortunately things got a bit out of hand, and before long she was being subjected to some lewd propositions. I finally had to confiscate the radio before we got into serious trouble from the paving crew as well as the now-defunct Department of Communications, which frowned on people fooling around on radio channels.

The Forest Service was unaware that I had their frequencies programmed into my truck radio, as well as my handheld unit. I thought it better not to let them know—that way I could hear what they were saying about my slashburning operations, and me in particular. This proved to be useful on a few occasions, notably on that escape that took place on the summer burn at Marl Creek. The toughest part about listening in on those conversations was resisting the overwhelming urge to pick up the microphone and add a few choice words of my own. I managed to avoid temptation but ended up with a few holes in my lip in the process.

•

Communication breakdowns could lead to really serious consequences, particularly when explosives were involved. During mop-up operations on an escaped slashburn at the top end of Bush River, we had to pump water a long way up the mountainside. A Wajax fire pump can only pump water so far uphill before the pressure at the nozzle end is insufficient to be effective, so a relay pumping set-up is used in these situations. The first pump moves water to a relay tank, or sump, where a second pump is set up to pump the water further up the mountainside.

Rather than having a portable tank assembly packed up the steep slope, I decided to blast sump holes and line them with plastic. As there were air operations taking place on other parts of the fire, I radioed the Forest Service fire camp further down the valley to inform them that I would be blasting, and I ordered them to keep their helicopter from flying anywhere near the site. I told them specifically that I'd call them when I was finished to give them the all-clear. I set off one charge and was preparing another one higher up the mountainside when a helicopter suddenly appeared overhead. I radioed the pilot directly and asked him what the hell he was doing in my no-fly zone. He told me that he'd been given clearance from the FS fire camp, so I got hold of them next. Those idiots told me that they'd heard the first blast and had assumed I had finished the blasting operations. I gave them royal shit, as did the pilot when he landed. Otherwise, the operation went well, and we ended up with some excellent relay sumps that you could lie in to cool off when the pumps weren't running.

Face-to-face verbal communication with the Forest Service was a different matter altogether, as my relationship with the government ministry was not the greatest. The procedure that I'd inherited from my predecessor was to beat them up verbally whenever there was a confrontation and, if the situation warranted it, threaten to continue the process physically. While most forest officers were reasonable to deal with, there were one or two who seemed to have spent many years in self-abusive relationships. The arguments usually stemmed from discussions about how my slashburn escapes were being fought. Whenever I felt that I was being pressured into carrying out an operation that I considered hazardous to men and machinery, I'd dig my toes in, particularly when I was told that we should be attacking a fire on a steep slope at night with heavy equipment. To my mind no patch of timber was worth the risk of rolling a Cat down a mountainside, and this argument degenerated in one instance to the point where I finally invited the forest officer concerned into the parking lot so I could sort him out. I won out eventually (in his office, not the parking lot), but it finally got to the point where I wasn't allowed to go to meet with the Forest Service on my own if a major row was forecast. My supervisor would accompany me in order to restrain my outbursts, although I'm afraid he didn't have a lot of time for the Forest Service either.

Not all the arguments took place within the office environment, as there were some good ones that went on out in the woods. Here there seemed to be more opportunities for contention, and to my great surprise I sometimes found myself arguing that a block should not be burned. I enjoyed burning very much, not because I'm a pyromaniac, but because I found it a challenging job and would therefore take every available opportunity to light up any of the blocks that were on the list to be burned. I did not, however, like the idea of burning a block where the fuels were too damp, as the result would be a patchy burn that didn't achieve the desired silvicultural objective of creating plantable ground.

This was the cause of a shouting match that took place out in the Beaverfoot on a joint field inspection involving the company and Forest Service staff from both the district and regional levels. The slash fuel on one particular block was not only too damp to burn, but the duff layer contained ice crystals—yet I was informed that it would be possible to light up that day. There was the usual shouting and yelling, mostly from me, and I think my supervisor dragged me off to the truck as always.

One joint site inspection trip was a little out of the ordinary, however. We stopped at several blocks in the course of the day, and for once there weren't too many disagreements. When we got back to town later, we heard that a fire had been reported on one of the blocks we'd visited, and the cause appeared to be a discarded cigarette butt. There was only one smoker in our party, and he was a forest officer, but to this day he denies it was his butt. It was a refreshing change: normally it was my butt that was in trouble on these tours.

In the forest industry, receiving an official written instruction from a forest officer is rather similar to receiving a traffic ticket from a police officer, and equally unpopular. The instructions are written on what's known as a 242 form, and some years I'd accumulate quite a collection of them by the end of the burning season. I'd have to sign for each one to acknowledge receipt, and I would do so readily just so the issuer of the 242 would go away. This wasn't the case in one instance when I disagreed so strenuously with the instruction being issued that I flat out refused to sign. The forest officer refused to leave my presence until he got a signature, so the situation turned into a Mexican standoff. Eventually we reached a compromise: he wrote up another 242 stating that I'd refused to sign the first one. Once I'd signed this second scrap of paper, he went away reasonably happy and I added my copy to my collection of official toilet paper.

The Forest Service did an audit in 1986 on a couple of our slashburns that had resulted in a little over-achievement. It was fair enough, as they'd both required the services of the air tankers, which are a provincial resource and rather costly. I was accompanied by two of the Woodlands staff, who sat on either side of me so they could kick me under the table if I got verbally offensive. During the audit I was asked how it would be possible to avoid escapes on similar burns in the future, one of which had taken place in the spring in Moose Creek and the other in August at Marl Creek. I suggested the only way to completely avoid any possibility of escape was to avoid burning at those times of the year, as weather is unpredictable, at least when it comes to reliable wind forecasts in that part of the country. Before my legs received too many bruises, I managed to rub it in that the Forest Service had had their own escape on a block they'd burned that spring. Their burn also escaped due to sudden, unexpected high winds, and it had nearly burned up the company's buildings at the log dewatering site at Kinbasket Lake. If they did have an internal audit on their own misfortune, I never got an invitation to sit in. Sometimes I got the distinct impression that as far as my standing with certain Forest Service staff went, I was about as popular as cat shit under the bed.

As a consequence of the audits we were informed that in future the air tankers might not attend our escapes if they should be needed on fires that had higher priorities—in other words, those endangering life and property nearer to civilization. I didn't like having to call for aerial assistance, but sometimes it was necessary. To cover ourselves, from then on I added a clause to some burning permit applications that stated we'd be calling for air tanker support should there be an escape that was beyond the immediate capabilities of the resources at our disposal.

I wasn't the only one in the company to have communication problems with the Forest Service. One of the logging supervisors took action on a wildfire up the Sullivan River with two bulldozers diverted from a nearby road construction project, even though the fire was the responsibility of the Forest Service. He started them building guard around the fire in order to try and minimize the spread. The Forest Service turned up in force and immediately redirected the Cats to brush out an old fire road so that the fire camp trailers they'd ordered wouldn't get their paint scratched as they were hauled into position. It seems they were more concerned about their creature comforts than attacking the fire, which they seemed to think would take quite some time to put out. The logging supervisor was so pissed off that he washed his hands of it all and walked off the site.

For some reason the Forest Service seemed to have different priorities when it came to attacking fires, or maybe it was something to do with miscommunication. Somehow I ended up in charge of extinguishing a small fire up the lake at Trident Creek that was again the responsibility of the Forest Service. I was waiting for men and equipment to be flown in from another of their fires that was being wound down on the other side of the lake.

The first helicopter sling load to arrive was full of hard hats for some reason. I wondered at the time if they were planning on sending a sling load of matching personnel on the next trip, but eventually a fire pump was flown over, along with four young, inexperienced firefighters. Unfortunately they hadn't thought to send a nozzle along with the pump, as I discovered when I found one of the firefighters holding the end of the hose in bemusement— the water stream only went out as far as his boots.

The Forest Service seemed to use helicopters less than cost-effectively, which occurs on some wildfire control operations to this day. A fine example was on a wildfire west of town, where they used a helicopter to fly crews to the top of the fire each day, which was a complete waste of money, as there had been a road constructed to the top. I heard about it from a couple of Cat operators, who told me they'd drive up to their machines in the morning, drink coffee as the Cats were warming up and watch the helicopter drop firefighters off at the adjacent helipad.

The Forest Service had the same type of electronic weather stations as the company, the only difference being that theirs had telemetry, which meant they'd get daily weather readings sent in automatically over their radio network. The company didn't have the money for this luxury, so I'd have to drive out to the weather stations periodically and attach a battery-operated printer to extract the readings. I just couldn't resist the temptation to screw up the Forest Service in a small way, so I made a tape recording of the data stream as it was transmitted over their radio system one day. The following day I played it back into the microphone of my truck radio at the same time as their new data stream was being transmitted. My radio was more powerful than the ones connected to their weather stations, so my phony information overrode theirs. I did this for a day or two until I got tired of the game. I don't think they ever figured out why the weather had suddenly become so consistent.

This was a bit more sophisticated than what I'd done to them a bit earlier when they were fighting a fire up Bush River and were using our logging camp for crew accommodation. They were also using our company radio channel for ordering supplies and such. I'd amuse myself by jamming their transmissions right in the middle of a long grocery order, so it had to be repeated. Too bad I don't have a talent for imitating voices; I could have really screwed things up for them.

My differences of opinion with the Forest Service were no secret from the burning crew, since they'd often be present when words were exchanged out in the woods. They got into the spirit of things, to the point where I watched two of them attend to an unpopular forest officer's pickup truck: one of them let air out of a couple of tires while the other was busily stuffing a tomato into the fuel tank. I hadn't incited them to commit this act of sabotage, but I didn't feel inclined to put a stop to it either. Besides, I'd done something equally bad when another unpopular forest officer came up to inspect a burn we were conducting on a steep side hill one afternoon. I made him leave his truck at a lower landing and ride up with me, telling him that there was too much congestion at the top landing. While he was parking his truck, I took a stick of dynamite out of the magazine, cut it open and smeared the stuff on the passenger door handle of my truck before he climbed in with me. I suspect he never did figure out what caused the headache he would have inevitably ended up with. The dynamite contained nitroglycerine, which will give you one hell of a headache when it's absorbed through your skin, as I know from experience.

There were a lot of good forest officers out there, nonetheless, and perhaps there were a few small flaws in my personality that didn't help me to get on with the remainder.

•

Miscommunication (or rather, lack of communication) within the company could have amusing consequences. Each season I would make up a list of blocks that were to be burned, after discussion with the silviculture department. Once broadcast burning operations began, I'd strike them off one at a time when they were completed and consult the list to see what could be lit up next. While flying back to town one afternoon with the helitorch slung underneath the helicopter and a forest officer sitting conveniently in the back seat, I realized we were near a small block that was on my list. The forest officer happily wrote me out a burning permit right then and there while the pilot circled the block with his finger pressing the torch trigger. I signed the necessary document as the first burning blobs of napalm hit the slash below and congratulated myself on our efficiency. On my return to the office, however, there were no further congratulations: it seems the silviculture department had planted the block just days earlier, on the assumption that I wouldn't be getting around to burning such a small block. They had forgotten to tell me, so most of their little seedlings had gone up in smoke and had to be replaced.

The burning crew had no problem when it came to communicating their thoughts to me, even at a distance. While they were waiting for me to fly out and join them on one particular block we were going to burn, they'd written a message on a switchback in the road using napalm gel. When they heard the helicopter in the distance, they lit the napalm; unfortunately they got the timing wrong and it had mostly burned out by the time we arrived overhead, but it was the thought that counted. The message was

"Fuck Off Raeside," with some additional fine print further down the road that I couldn't quite make out. Which was probably a good thing.

●

Unfortunate fact:
  Radio waves travel at approximately 186,000 miles per second, which means that news of your burn getting away will invariably reach town long before you do.

10

BETWEEN FIRES

Not all the work I did during the time I was with Evans Forest Products involved lighting or fighting fires. I went out timber cruising while I was part of the IFFS crew based in Golden during the 1977 fire season. As a compassman, my job was to walk ahead of the timber cruiser on a designated bearing with the lead end of the plastic measuring chain attached to my belt. This meant that I'd be the first to get to any patch of ripe huckleberries that happened to be growing in our path. The downside was that I also got to find underground hornet's nests first, but if I was quick on my feet, I could get the insects stirred up and then let the guy following take the blame. I wasn't the most popular compassman due to my habit of never deviating from the compass bearing, no matter what was in the way. If a cliff was in the way, I'd go straight up it rather than do an offset bearing, much to the annoyance of the man on the tail end of the chain who'd have to follow behind and dodge any loose rocks I might have dislodged while climbing.

I went on reconnaissance timber cruises in a few of the valleys adjacent to Kinbasket Lake during the following two summers and got to see the western side of the Rocky Mountains up close. While working up the Sullivan River, we had to tow two of the crew across the river at the end of the day, as meltwater had caused the river to rise so that it was too deep to wade. We managed to get a rope across and then hauled them one at a time through the ice-cold water. It was a bit like hauling in a very large fish, except fish don't use that kind of language once you've got them out on the bank.

One reconnaissance cruise in a side-drainage of the Sullivan was especially challenging because for some reason I had to pace the distance as I followed the bearing. Pacing on flat, open ground is easy, but inaccuracy increases sharply as the terrain becomes steep. It gets even worse when dense undergrowth and creeks are encountered, particularly when you fall off the log you're balancing on while crossing a creek. After fishing yourself out and climbing back on the log, you have to remember what the pace count was at the moment you fell in. I ended up pacing a total distance of well over three miles uphill and down as well as falling into a couple of creeks. I needn't have worried about accuracy as it turned out, as it seems my cruiser had misread the scale of the map he'd been using.

We all enjoyed our time in the fly camps, as we were being paid to be in some really spectacular Rocky

Mountain settings. There was one individual who didn't quite fit in, however, as he seemed much too fastidious to be a timber cruiser. We realized this fact when he insisted on purchasing dishwashing liquid and a draining rack for washing up the utensils after each meal. The rest of us would simply scour the pots and plates with silt scooped out of the riverbed, and we regarded him as a bit of a freak. Our suspicions were confirmed when he was spotted washing his face at the river's edge in the morning, with a sponge bag placed on the ground beside him. It came as no surprise to anyone when later in the season his head came into contact with his cruising partner's clenched fist after a heated discussion.

•

We were issued pocket flare launchers along with a supply of flares. The red flares were supposed to be used for attracting the attention of a helicopter pilot if he was unable to spot us on the ground. The others were "bear flares," which were designed to explode with a loud bang in order to scare off unfriendly bears. They weren't much of a deterrent, as I discovered when testing one on a couple of grizzlies near Mica Creek. One ran a few feet and then stopped, but the other refused to move. They were more effective on humans, and whenever two crews were camped on opposite sides of a river, the first to get up in the morning would fire a bear flare at the other camp as a wake-up call.

We went through a lot of flares during the field season, particularly if there'd been a flare battle between crews. It did get a little out of hand when we were timber cruising up the Wood River. Individual crews had been dropped off by boat and were being picked up again at the end of the day. As the boat came close to where one crew was waiting, they fired a red flare directly at us. It passed just above the boat, so naturally we returned fire, switching to bear flares when we ran out of the red ones. As the battle heated up, we suddenly realized there were several five-gallon jerry cans of gasoline sitting on the exposed deck, all leaking fumes. It would have been interesting if a flare had landed right amongst them.

One of the cruisers had a bad habit of carrying his flare launcher in a pocket of his field vest, loaded for bears and cocked ready to fire. I'd warned him that it could go off if it got snagged on a branch, but he didn't believe me. He changed his mind the day it did go off accidentally: the flare hit his ear and then angled off into the air before exploding, luckily for him.

I spent some days staying in a tent pitched beside the Cummins River while we were timber cruising in that valley, part of which is now in a newly created provincial park. Why anyone would want to venture into the Cummins Valley for recreational purposes is beyond me; it's a mosquito-ridden hellhole with swampy ground and buckbrush that made travel slow and unpleasant. We were there in September when the river was low enough to provide dry ground for pitching a tent, but when it rained during our stay, we woke up one morning to find the water had risen to the point where the tent was surrounded. I always took dry kindling into the tent with me at night, along with my boots in case it rained, but this time it didn't help—there wasn't even enough dry land to make a fire for cooking breakfast.

    The company eventually set up a logging camp near Tsar Creek, very close to where we'd lost our beer cache a few years earlier. I was given the job of supervising the setting up of the trailers, which were sent up the lake by barge and then towed up the hill to the site by a small tracked loader. The trailers were familiar; I'd helped remove a few of them from our other camp at Gold River further up the lake. The occasion had been memorable, as two of the three-man hauling crew sent to move the trailers were drunk when they arrived and nearly got into a fight with the camp cook. After that they started fighting amongst themselves. I helped set wheels under the bunkhouse units, and it was a bit unnerving to work underneath them as they were being jerked erratically by the drunken tow truck operator who was yelling abuse at us from above.

    I helped out on silviculture regeneration surveys for a brief time—very brief, in fact, as it seemed I didn't have the right temperament for the job. The survey involved the counting of little conifer seedlings that had sprouted naturally in areas that had recently been logged over. It could be extremely tedious when there were a large number of them growing within the radius of the sample plots.

The plots were divided into four quadrants, and I seem to recall that there only needed to be a minimum of four seedlings in each quadrant for the plot to be considered fully stocked. It didn't take me long to decide that there was absolutely no point in counting anything over the minimum, so I started weeding out the surplus. As I was in the lead, dragging the plastic chain along on a compass bearing in order to establish the plots' centres, I'd have time to do it, so by the time the silviculture surveyor caught up with me, there'd be uprooted seedlings strewn all around.

Nobody had ever done anything like that on a regeneration survey before, but as I pointed out, the procedure manual didn't specifically say you couldn't weed the plots. As long as they were technically stocked, why not make the job a lot easier? It actually did get a lot easier for me, as I wasn't asked to help with those tedious surveys ever again. However, I did spend a week doing mountain pine beetle surveys near Castlegar at the end of the first fire season with IFFS. We only found one tree that showed visible evidence of pine beetle damage, but that was back in 1977 when the epidemic was just getting started. The little pests have chewed their way through a lot of BC forest since then.

I had no further involvement with beetles until several years later, when I was invited to join a field trip with Evans Forest Service staff to see what could be done to halt mountain pine beetle spread in the Golden Forest District. The key attendee was a forest pathologist from the FS, and I discovered that I'd been included because of my blasting experience. The pathologist had an idea that it might be feasible to wrap detonating cord around the trunks of infested pine trees and kill any beetles living under the bark with the resulting explosion. It was an intriguing idea, and rather tempting, but the same result could have been achieved more cost-effectively by removing the bark with an axe. Not surprisingly, the proposal was rejected after a very brief discussion.

•

While I was setting up the camp at Tsar Creek, I somehow ended up with the additional job of supervising construction on some logging spur roads. It went fairly well, apart from an unusual number of equipment breakdowns. There was the odd interesting moment, however—one in particular was the day I went out to help the crew clear a suitable site for developing a gravel pit. I ended up hooking chokers to the trees being cut down by the faller, which went well until one of them headed straight for the Cat as I was walking around the front of it. I never saw the tree coming, but I heard the operator yelling a frantic warning and realized instantly what was about to happen. There wasn't enough time to climb onto the tracks and get inside the ROPS canopy with him, so I ended up pressing myself as tightly as possible against the undercarriage on the lee side of the machine. There was a loud bang as the tree struck the Cat and broke into two pieces, and from my sheltered position in with the track rollers and sticky mud, I saw the top section crash to the ground, with the broken end landing a few inches from me. The faller up on the hillside above us was laughing for some unknown reason, but the Cat operator and I failed to see the joke.

I occasionally did one or two small blasting jobs on the logging roads, just to help out. A large boulder had rolled onto the road just up the valley from the Bush River camp, where it was partially blocking the road, so while pickups could get past, there was no room for logging trucks. I went there in the evening with a couple of the burning crew to do our good deed for the day by getting rid of the offending rock with explosives. After sizing up the problem, I decided to place a bag of Amex explosive on each side of the boulder, connected together with detonating cord so they'd both go off simultaneously. I figured that the two shock waves would meet in the centre of the boulder and shatter it into pieces that we could easily roll off the road. As usual, we wanted to get a good look at what happened when the blast went off, so we crouched expectantly in the ditch a little way down the road. There was a very loud bang, and the trees across the road from the blast looked like they'd been hit by a Category 3 hurricane. The next thing we saw was a lethal charge of fly rock heading along the road in our direction, which forced us to lie as flat as we could in the ditch as it whistled overhead. Once it was safe to stand up, we went to see the result, which turned out to be what might best be described as overachievement. The boulder was in smaller pieces than I had expected; in fact, it was difficult to find any pieces larger than a grapefruit. Not only had it been blown to smithereens, but part of the road had been blown away as well. We ended up scraping jagged shards of rock off the road with an uprooted mile-post sign so we wouldn't get complaints about blown tires from angry truckers. We then had to fill the crater in the road with gravel and stamp it down as much as possible so the road foreman wouldn't be asking awkward questions.

I was asked to remove some of a bedrock outcrop that was protruding above the road surface on one of the company's logging roads, which had been causing problems for the grader when it passed over it. The easiest way to do this was to do a mud-cap blast, which involved placing a stick of dynamite on the rock outcrop and then covering it with mud. The mud directs the blast downward just fractionally long enough to shatter the rock. There was a logging crew working immediately below me on the same mountainside, so I called them on the radio just before lighting the fuse to warn them that they might want to take cover when they heard the sound of my air horn, as they should expect a shower of rock fragments. I never heard any complaints after the blast, so presumably nobody got hit with any of the pieces and all their windshields were intact. Blasting close to logging operations isn't always popular for another reason: pieces of fly rock can get embedded in the trunks of trees and create havoc when a power saw chain strikes one.

When it became very dry in the summer, I'd be out keeping a close eye on the conditions in the woods, making sure that logging operations were in compliance with the fire prevention regulations. When the fire danger reached a certain point, operations were put on "early shift," which required them to be shut down at one o'clock in the afternoon, but sometimes it seemed that people's watches were running a little slow... It didn't always matter, depending on the site conditions. Cat logging in a hemlock stand on boggy ground posed far less of a fire hazard than high-lead yarding of dry fir logs on rocky ground on a south-facing hillside. There were times that I'd shake my head though, as I did when I found one haywire contractor doing some welding on a broken-down skidder right in the middle of dry slash on a blazing hot afternoon in the height of summer during a shutdown period. When I pointed out to him that perhaps it really wasn't such a good idea, he pointed to a small bucket of water standing nearby, saying that he was well equipped to deal with a fire in the unlikely event that one should start. I don't think he really understood why I insisted he turn off his welding torch at once.

The logging crews didn't like being put on early shift when the fire hazard was high—it meant getting up very early in the morning and sitting around camp in the afternoon with little to do. They did understand the reason for the restrictions however, and had no desire to be the cause of a fire, seeing as they'd be the ones who'd have to go and put it out. One high-lead crew was less cooperative than the rest though, as I found out when I showed up unexpectedly while they were logging near Tsar Creek one hot afternoon when they should have been shut down for early shift. It was three o'clock, and they were yarding dry fir logs downhill on a steep sunbaked hillside where the slash was so dry that it was only a matter of time before the friction of a couple of logs rubbing together would send the whole place up in smoke. After watching for a few minutes, I went up to the yarder to have a chat with the operator. When he saw me approaching, he shut the machine down, whereupon the rest of the crew headed down to us. "So, you're the asshole who's going to shut us down and cost us money!" one of them began accusingly. I informed them that actually I was the individual who was going to sit on a stump and take some great photos of fire crowning up the mountainside after the inevitable happened, should they continue working. And that they were the assholes who'd spend the next week or two putting it out. The crew grudgingly packed it in for the day.

•

I did quite a lot of flying in helicopters; it was generally the fastest way of getting where you needed to go in the mountainous terrain. It certainly made timber cruising a lot easier when you were set down on a landing spot above the treeline in the morning, since from then on it was all downhill. Sometimes the landing spot would be a clearing in the forest just large enough to accommodate the helicopter's main rotor blades with a bit to spare and a log crib on the ground to provide the machine with a reasonably level place to set down. They weren't always constructed to a high standard, as we discovered when a helicopter tried to lift off again but got one of its "bear paws" caught on a helipad crib log that hadn't been properly trimmed. A bear paw is a piece of plywood that is attached to the helicopter skid and is designed to prevent it from sinking when landing on soft ground or snow, a bit like a snowshoe.

My timber cruiser and I were crouched down on either side of the crib, as there was nowhere else to go: the helispot clearing had been hacked out of the forest, and the undergrowth was fairly dense. (It was always safer to stay right beside the helicopter after disembarking until it lifted off again, because moving elsewhere could be fatal: you could walk back into the tail rotor and lose an arm, or walk uphill and lose your head due to the main rotor.) The pilot increased power in an attempt to break free, and I was interested to notice how the laminations of the bear paw plywood were separating under the strain. At this point he should have set back down and moved sideways, or perhaps shut down and looked to see what the problem was, but for some reason he kept trying to lift off. All that did was cause the machine to pivot to one side as it lifted unevenly, and the main rotor blade started to get closer to the ground on the side I was on. There was nothing I could do except try to make myself as small as possible and hope he got unhooked before the blades struck the ground and the machine became a giant Cuisinart. Finally the bear paw broke free and he took off. I don't think the pilot enjoyed the experience any more than the two of us on the ground did.

The helicopters we flew in were very reliable, but now and then there were a few minor mechanical problems. One winter morning I went down to the airport to fly out to Gold River camp in order to shovel snow off the flat roofs, and I found the pilot hitting an igniter module on the engine with a hammer, as it "had been giving trouble lately." I believe the type of helicopter we were flying in has two of these devices, but I was glad that a different machine flew out to pick us up at the end of the day. On the trip back to town I was sitting next to the pilot, and I happened to notice that the engine chip light was on. This light will go on if any metal fragments are detected in the lubricating oil passing by a sensor, and generally it means you should land immediately before the engine becomes completely dysfunctional. The pilot didn't seem to have noticed. I've never liked being a back-seat driver (or front seat in this case), but as I had a direct interest in the engine's well-being, I felt I had to point out the light to him. He replied that the light had being coming on intermittently for a couple of days, and the problem seemed to be a faulty sensor. He then added: "...but just in case," and dropped our elevation several hundred feet until we were flying quite low above the traffic on the Trans-Canada Highway.

My accommodation in town varied over the years and wasn't always that inviting, which is why I tried to find an excuse to stay in company logging camps whenever I could. Now and then the entire burning crew would be based out of camp, but I'm afraid we weren't always ideal guests.

When we were burning, we'd often work quite late, sometimes not getting back into camp until ten o'clock at night. This didn't make us too popular with the cooks, as they'd have to leave food out for us in the steam tables. In addition, we often pissed off any loggers who happened to be sharing our bunkhouse, as the burning crew liked to relax with a few drinks after their late supper. Often they'd get a little noisy and I'd have to try and shut them up before some large, ornery logger came out of his room to knock their heads together. One night when I was attempting to get the crew to tone it down a bit, one of them came at me with a knife, but he wasn't too steady on his feet, so it was fairly easy to take it off him and confiscate it until morning.

The other inhabitants of the bunkhouse would get their revenge in the morning, as they'd get up at the crack of dawn, while we didn't have to crawl out of bed for another hour or two. The gravel crew would be up first, and they'd kick the doors open on their way out, followed by the loading crew a bit later, who had a go at the doors as well. The loggers would be the last ones to leave the bunkhouse with a few final door slams.

I was in a bunkhouse at Gold River camp when the camp manager rushed in to tell me that there'd been an accident a few miles up the road. I jumped into my truck and headed off as he followed with the company ambulance. When I got to the scene I saw a mangled pickup truck quite a way down the mountainside below the road, with a couple of rescuers already there. I took the basket stretcher down with me, but from the distance down the slope and the condition of the truck, I expected the driver might be dead. When I got to the wreck, I was amazed to see him bloodstained but alive. We put him in the stretcher and hauled him back up to the road. It wasn't easy, as it had rained and the ground was steep and slippery. By this time a helicopter had arrived and was waiting at a wide spot further along the road, so we loaded the stretcher into the ambulance and drove the patient to the helicopter, whereupon he was whisked off to the local hospital for repairs.

The truck belonged to a logging contractor and was being driven by one of his men, who had been delivering a large replacement driveshaft to where the crew was working, which I discovered later that morning when the contractor came out to view the remains of his pickup. He was more concerned with getting the driveshaft to its destination than with the condition of truck and driver, and I had to go back down the hillside with him to help him pack the damn thing back up to the road. As it turned out the driver had received only minor injuries, and he probably could have climbed back up to the road unassisted if he'd been left to his own devices.

•

None of the burning crew ever received an injury of any significance while I was in charge of burning, which still seems little short of a miracle considering some the dangerous conditions we worked in. Burned-out snags, rolling rocks, steep ground and excessive speed on gravel roads were some of the natural hazards we faced. There were a few unnatural hazards as well, but I won't go into them.

We had numerous close calls, however, such as the time we were lighting up a steep high-lead block when debris started rolling down the hillside, including a sizeable rock that was headed straight for one of the crew. We watched with interest as he ran away from it straight downhill, in the same direction the rock was headed, instead of just moving across the slope to get out of its path. It was almost like a cartoon sequence, but I don't think he found it amusing. I nearly got nailed a day or so later on the same block when a log started rolling down the slope toward me. I had no time to get clear, so all I could do was dive to the ground behind a stump in an attempt to protect my head. At the last moment the log struck something, turned ninety degrees and slid right past the stump I was hiding behind.

There were minor mishaps, like the time I was backing up a rough spur road with my head out the window of the truck. One of the wheels hit a rock and I bashed my head against the edge of the glass, resulting in a slight head injury. Back at the Woodlands office, one of the silviculture staff saw the blood and asked what had happened. When I told him, he led me outside to show me the circular shattering on the passenger side of his truck's windshield, which had been caused by the impact of a tree planter's head. They'd been driving out in the woods earlier that day when their vehicle had also encountered something solid. The tree planter had apparently been none the worse for wear and had shown no signs of concussion, but then again, in those days it was hard to tell with tree planters...

We didn't let minor injuries stop us from going to work, as nobody wanted to miss out on the fun. A senior member of the burning crew failed to show up on time for work one morning. Apparently he'd been helping drink the Texas mickey another of the crew had won at one of the bars in town, and in the process somehow his head had come into contact with a window. I didn't expect that he'd be coming with us that day, due to the combined effects of the accident and the alcohol, but surprisingly he did show up at the last moment, head bandaged, to take his usual seat in the crewcab.

One of the crew was a particularly keen hunter, and he started packing his rifle along with him during light-up. I had to tell him to leave it in the truck in case he mistook the largest member of our party for a bear as the light faded. During a meal break the hunter hung his pocket watch on a stump, then fired a round through it, though we were never quite sure why. I kept what was left of the watch and one day took it to a watchmaker for a repair estimate. He nearly fell off his stool when he examined it; I believe he thought there was a body out there somewhere with a matching bullet hole.

This same individual was sent across Bush River with another of the crew to retrieve the equipment from mop-up operations on the block that had been ignited by accidental pilot error. They had an aluminum dinghy with them to get across, as the water was running deep and cold. When I drove past that location later in the morning, I happened to look across in time to see the two of them fighting in the middle of the river and the boat being carried off downstream on its own. The bigger man appeared to be trying to drown the smaller one, who understandably wasn't cooperating. I decided I'd prefer not to get involved in their dispute, so I continued on my way. The boat was eventually rescued from where it had washed up on the bank, but I have a suspicion that at least one length of hose was lost overboard during the altercation.

The crew wasn't always regarded favourably by the company's upper management, I'm afraid. When they were sitting in the crewcab one morning, with the usual wreaths of smoke coming out the windows, the general manager walked past on his way from the main office to the Woodlands office. After noticing what was going on inside the vehicle, he stormed through the door in a fury. "Who are those degenerates in that truck outside?" He was informed that they were in fact the slashburning crew. "You mean they actually work for us? Make them park around the back from now on—they smell bad!" If he'd known a bit about the crew and what they'd done from time to time, he'd have probably fired them all on the spot.

A number of odd incidents involving the burning crew took place, but they are probably best forgotten. One that does spring to mind, however, was the time the crew had forgotten to fuel up the crewcab and drove around town at night until the engine quit. Since there was a slip-on tank full of fuel in the back of the truck, they simply refuelled the truck and carried on. Unfortunately they'd forgotten that the fuel was the mixture of diesel and gasoline we used for pile burning, as it was late in the season, and after a short distance the engine quit again. But they were in a hurry to get to the pub, so they simply abandoned the truck in the middle of the main street with the keys in the ignition and walked the rest of the way. I received a call early the next morning from the police, informing me that they'd like the offending vehicle removed, as it was impeding traffic. Fortunately nobody had stolen any of the equipment out of the back of the truck overnight— there were a few thousand dollars' worth of saws, fire pumps and other tools. For some reason the crew seemed to think that the verbal scourging they received when I finally caught up with them was unjustified.

•

Helpful fact:

Contrary to popular belief, the porcupine cannot shoot its quills at an attacker. Some of them will fall out, however, as it's being prepared for cooking.

11
PILE BURNING

Wet weather would put an end to broadcast burning by late fall, and it would finally be time to relax a bit. The last of the mop-up was finished, and we could get started on burning landing debris piles, as by this time of the year there'd be minimal chance of them spreading into the nearby forest. Burning slash when it had been piled up was a lot easier than when it was spread out all over a cutblock. It was just a matter of driving up, pouring on some fuel and then tossing in a lighted match.

Sometimes it wasn't possible to get the truck close to the piles if the road had been washed out or was not passable for some other reason. In these instances we'd have to walk the rest of the way, carrying the fuel in five-gallon pails. One of the crew had a bad habit of quietly emptying fuel out of his pail in order to make it lighter, so I liked to walk behind in order to keep an eye on him. The pails were discarded oil containers that could be found lying around in quantity near logging sites in those days, as the preferred method of recycling was to toss them into the bushes or over the downhill side of the landing.

The procedure for lighting a pile once the fuel had been poured into a suitable dry spot at the base was to strike a match, throw it on and turn so your back was to the pile. This way you wouldn't lose your eyebrows when the match landed in the fuel, which was the usual mix of diesel and gasoline. If it was a particularly cold day and there was a little too much gasoline, the resulting explosion could lift you off your feet. It happened to me numerous times and I got quite used to it—the feeling was as if a giant hand had hit you in the back.

•

We preferred using strike-anywhere matches, as they could be struck with one hand. I used to strike them on my zipper, until the plastic eventually became fused, at which point I switched to striking them with my thumbnail. I was in the habit of carrying a handful of loose matches in my shirt pocket until the day they ignited from rubbing together, at which point I switched to carrying them in their box in my jacket pocket. This was much safer, as when they eventually ignited from friction in the same manner, the box contained the flames long enough for me to remove it from my pocket before they ignited my jacket.

Some of the piles were quite high, and we'd have to climb down the steep face in order to find a spot to pour in fuel. We'd end up lighting in several places along the pile, which meant climbing around on slippery logs as the first fires were taking hold. It could be a bit unnerving when the pile started shifting, and our greatest fear was that it would suddenly move under our feet and someone would get trapped between large logs in a burning landing and end up getting roasted like Joan of Arc. I made up chemical delay igniters that would give us more time to get further away before they burst into flame. They gave good results on piles and worked equally well when I dropped an activated igniter into the open pail of fuel one of the crew was carrying. (Carrying volatile fuel in open pails near a fire wasn't a particularly safe practice and was done only when a fire looked like it needed a little bit of a boost.) We both watched as the small plastic cylinder began to emit purple smoke and then started going in circles on the surface of the liquid, rather like a miniature motorboat. A few seconds before it burst into flames, the holder of the pail had the presence of mind to empty it, fortunately not in my direction.

The company had a very large deck of old hemlock logs that had been sitting on their timber holdings near Revelstoke for a few years. They'd finally given up trying to find a buyer for the logs, so I was instructed to go over there and dispose of them. We had quite a job getting the deck burning, but we finally managed with the help of a load of old tires from the local dump. This was in the days when no one cared what you burned; I don't know how many pickup loads of tires we collected from the tire shops to get wet landings going late in the burning season. Nowadays the Ministry of Environment would come down on you like a ton of bricks for burning such things, but back then we never heard any complaints. It's amazing how hot a tire will burn once you've got it started. One day we discovered a large skidder tire that had been discarded at the side of an inactive logging road and just had to stop and light it up. The resulting column of thick black smoke could be seen for miles. One of our favourite diversions was to pour a gallon of burning fuel inside a twenty-inch truck tire, light it, then set it rolling off down a steep hillside. Of course, this would be done late in the year when there was no danger of starting a forest fire. It was particularly spectacular when done at night, as the tire would shoot out flame as it gathered speed downhill and bounced in the air whenever it hit a rock.

A couple of weeks later I was asked if I'd burned the hemlock deck, and I was able to report that it had been reduced to ashes, expecting that the company might be pleased. To my surprise I was told that they'd finally found a market for hemlock logs, and they were hoping that maybe I hadn't got around to the job.

The next time I was ordered to dispose of a surplus log deck on a landing, I made very sure that they weren't likely to change their mind afterwards. It was a deck of short logs that was, in the company's view, not worth the cost of loading onto a truck and hauling out to the mill, so they wanted it burned. I had a suspicion the burn wouldn't be too popular with the Forest Service, as they'd regard it as a waste of useable wood, but I followed the order anyway.

Sure enough, just after we'd started lighting the short deck, we noticed a green FS pickup down in the valley below, heading rapidly in our direction. When they got up to the site, they were not at all pleased to see what we were doing, as they'd come to do a waste-assessment survey of the log deck. I told them to go right ahead, adding that they'd better work fast, as fire was spreading rapidly through it, but for some reason they declined. I told them that I was just following orders, and that if I'd known of their mission, I wouldn't have dreamt of sabotaging it—but knowing my record of cooperation with the Forest Service, they didn't believe a word of it.

We didn't endear ourselves to a pair of Albertan hunters either one day, as we let them drive past us on a dead-end logging road, then promptly lit up a large landing pile right next to the road. The heat that was generated once the pile was fully ablaze would have burned the paint off their truck—that is, if they could have driven back past it at all. It's likely they had to wait until it burned down, which would have taken a while.

It wasn't just people that sometimes got upset with what we were burning. While two of us were lighting up landing piles on Blackwater Ridge, a black bear suddenly burst out of a hole in the pile I was working on. It came so close that it banged into the fuel pail I was holding and then charged off downhill in the direction of the next pile we would be lighting up. As a precaution, my partner fetched his rifle from the truck and covered me while I lit the remaining piles, but the bear had obviously decided to keep right on going.

I can sympathize with the bear's agitation, as I'm sure it can be a bit of a shock when the location that you'd picked for a nice winter's hibernation suddenly goes up in smoke, and I was glad it had escaped in time. From then on, I'd peer inside any cavities at the base of a landing pile in case there was a creature of some type in danger of burning up. Some of my crew weren't so concerned, particularly the one who began lighting the base of a large pile while I was standing on the top. Mistakes like that can happen when it's starting to get dark in the evening, but in this case, when I yelled at him, he just smiled up at me and kept on lighting. Not a big deal, as the flames hadn't yet reached up to where I was standing, but I made a mental note to keep an eye on this individual in the future.

After working with fire for a while you get quite used to it, and we tended to play with it from time to time. During broadcast burning you might try to light around the man working next to you with your driptorch, so he'd be encircled by flame and be forced to step through it. Obviously you wouldn't do this in heavy slash fuels—the trick was to surprise your partner, not roast him.

I came up with a more interesting way of increasing someone's pulse rate when a few of us were hiking in along an abandoned spur road to burn piles. The road had sunk downward in places, and I surreptitiously poured fuel into one of these depressions that was directly in the path of one of the crew. I then laid a trail of fuel away from the hole and waited until the right moment before lighting it. My timing was perfect, for there was a fuel-vapour explosion beneath his feet just as he was stepping over the hole, one that definitely got his attention. Needless to say, I kept a wary eye on him for the next few days, as it was inevitable that there'd be some form of reprisal. I inadvertently played the same trick on myself once or twice while lighting landings that were covered with snow. It was hard to see exactly where the fuel I'd poured out was located, and it wasn't until I threw down a match that I discovered I was standing directly over it. The result was that I'd be in the middle of a blast of flame, which wasn't a big deal, as I'd learned by experience to hold my breath and close my eyes whenever I lit the match.

•

Sometimes we had to burn piles in locations that were inaccessible by road, and we'd end up going there by helicopter. This made the operation expensive, so we'd try and work as fast as possible, which led to a problem the time we had to burn a number of landing debris piles across the Columbia River, where a winter bridge had been removed. It seemed like a routine operation: fly over with fuel and land at each pile in turn, light it up as the helicopter waited for us, then fly off to the next one. Everything went like clockwork until we got to a fairly large pile and somehow managed to get our timing wrong. Two of us were on top of the pile, pouring on fuel, just as the third man was throwing a lit match into the fuel that he'd put on seconds earlier. The pilot said afterwards that the shock wave from the detonation had lifted the machine up off the skid on one side and also pushed in the bubble window partway. The rotor blades had been spinning at the time, so it could have been nasty if the helicopter had gone right over. Fortunately it dropped back down again, but the pilot was understandably upset, and from then on he set his machine down much further from the piles and made us walk back.

When money was tight, as it often seemed to be near the end of the burning season, we didn't have the luxury of air transport and therefore had to get to the piles some other way. One day, when there were some that needed to be burned on the wrong side of Bachelor Creek, I ended up stripping off all my clothes and wading into the icy water with my fuel pail in one hand and a box of matches in the other. I lit all the piles I could see and then waded back, but while I was warming myself up beside a fire, I discovered that I'd missed one—and had to plunge back into the water once again.

Inevitably explosives found their way into the pile-burning program, and sometimes I'd light a large pile by crawling into a cavity within the pile with a five-gallon pail of napalm wrapped in Primaflex detonating cord. When it went off, it would blast burning napalm through the logs and branches in the pile. I had a slight problem at one large pile when I'd crawled a fair way inside it to set one of these firebombs. After I'd lit the fuse, I turned around and started crawling back out, only to get my foot jammed in something. The fuse was the usual two-minute delay, but the crew didn't help my concentration by shouting in chorus that the time was nearly up as I was trying to free my foot. This was the same crew that stood on a spur road above me, watching as I carried yet another pail of napalm—complete with detonating cord, blasting cap and fuse, all ready to be lit—down the slope to set inside another pile. I suddenly heard laughter, accompanied by a strange roaring sound, and I turned to find that they'd poured burning fuel in an old tire, lit it and then rolled it down the hill aimed right at me. I had to run to get out of the way, still clutching my firebomb.

We managed to find a few other uses for napalm and detonating cord in the course of the burning season. While burning landings in the Glenogle Valley, we came across a discarded culvert that had one end crushed lying beside the logging road. For some reason I got the idea that it might make a good mortar, so we set it up on an angle against a pile of dirt, crushed end down, and I loaded a charge of Primaflex into it, followed by the usual five-gallon pail of napalm. Once the fuse was lit we hid behind the trucks, watching eagerly to see what would happen. When it went off it looked like a recoilless anti-tank gun, with a back-blast going out the crushed (and now mangled) end. We were unable to see where the pail of napalm landed, which was a bitter disappointment—we were all expecting some kind of fireball, since we'd attached a lit napalm-soaked rag to the lid.

I thought no more about this until years later, when I found out from my former supervisor that on the same day we'd fired this mortar, a logging truck had turned up at the log-yard scales with several of the logs on the trailer showing evidence of having been burned. To make it even more suspicious, there was apparently a lingering smell of napalm. I tried to find out exactly where the truck had been when it was hit, as I really wanted to figure out how far the projectile had travelled, but he wouldn't tell me. Reading between the lines, I suspect that my future with the company may have hung in the balance that day.

•

As the season began to wind down, the weather would get colder and it seemed that we required a bit of alcohol in our circulatory systems in order to keep going. A bottle of red wine would be shared at lunch, but now and then it would be a rather large bottle, or perhaps two if it was a really chilly day. One morning in the office, our boss asked us if we'd burned the landing piles on a certain cutblock on Blackwater Ridge yet. We replied that we didn't think we had, which might have made him slightly suspicious, and we added them to our to-do list for the day. When we arrived at that particular location, we discovered the piles had already been lit at least a day earlier, although neither I nor the other two in the truck with me could remember actually lighting them. We solved the mystery when we spotted an empty wine bottle lying nearby; it was the cheap Mountain Red brand we were drinking that season.

    Things got a little more out of hand later on when I went south to burn some landing piles near Invermere with the same two individuals. We'd agreed not to open our bottle of wine until we got at least ten miles down the road, but the cap ended up flying out the window before we'd even crossed the railway overpass on the outskirts of town. As a result we ended up having to stop at a liquor store en route to replenish our supplies before we stopped in at the Forest Service office to get the necessary burning permit. There was snow on the ground outside, and there would be a lot more of it where we would be burning, but the forest officer we had to deal with was a real worrywart. He droned on about us being required to supply two Cats and ten men in the event of an escape, and nervously peered out the window at our pickup sitting in the parking lot. All that was visible was a forty-five-gallon drum full of burning fuel chained to the tool box in the back of the truck and the third member of our party lying semi-comatose in the cab. "I suppose you've got all the necessary fire suppression equipment in the tool box?" he ventured. We nodded vigorously, hoping that he wouldn't come out to inspect. All that was in the tool box was an oily Jackall that didn't always work properly and a blunt Pulaski. I kept wishing he'd stop asking all these questions, as the refreshments we'd partaken of on the way down were catching up with me. He grudgingly wrote out the burning permit, which I signed with a wavery signature; then he asked if we knew where the piles we were going to burn were located. I'd forgotten the directions I'd been given back in town before I left by this point, so the forest officer drew up a map, or actually several maps, as he was using a small stationery pad to draw on. I leaned over to watch as he drew in roads, creeks and other useful bits of information, trying not to breathe on him. On the way out of the office, I dropped all the pieces of paper and managed to get them out of order when I picked them up.

The third man in our party was revived by fresh air after we'd driven down the road a mile or two, while I tried to make sense of the mixed-up instructions. They eventually got accidentally thrown out the window, and we had to rely on memory and dead reckoning. Eventually we found what we believed to be the piles we were supposed to light up and soon had them all burning merrily. Things got a bit weird from that point on, and I seem to recall that we decided to do a bit of curling on a frozen pond with empty wine bottles. They shattered before the game got well underway, and as we headed back to the truck, I had to rescue one of the party members who was hanging upside down from a barbed-wire fence and flapping his arms like a giant bat. The trip back to Golden was a bit rough and wasn't helped by the fact that we stopped at a pub along the way for a pick-me-up. I think I did all the driving that day, but I can't be sure. It would be a bit cramped on the seat of the pickup when the three of us were out burning, and we got into a routine whereby the man sitting in the middle would shift gears when asked to by the driver. Sometimes the middle man would change gears on his own initiative, if he thought the engine was labouring excessively, which would be hard on the gearbox if the driver didn't push the clutch down in time, so it led to occasional arguments. I never heard any feedback about the landings we burned that day, so they must have been the right ones, but we were never quite sure.

•

On rare occasions a burning landing pile would set fire to the adjacent forest, usually due to sparks being thrown by the wind. There normally wasn't any chance of a major conflagration since the piles were being lit so late in the year, but there was the odd exception. I once set fire to landing debris on a couple of high-lead blocks situated at high elevations, and I actually had difficulty getting to them due to snow on the access roads. After that I headed over to Revelstoke with the crew to do some pile burning there, and unbeknownst to me the weather improved during my absence to the point where the snow on the blocks disappeared, the slash dried out and the landing fires set the blocks on fire. On my return I was informed that the two escapes had burned right up the mountain until they ran out of vegetation, and that the Forest Service was a bit upset.

One of the more interesting escapes from a burning pile set fire to some hollow cedar trees situated in a swamp. We ended up going out at night to extinguish them, as there was the possibility that a tree might burn through and fall onto the adjacent main logging road as a truck was passing the following morning. It wasn't the easiest task to put the fires out, partly since it was pitch dark and partly because most of us had been out drinking when the call came in. We had only one flashlight with us, and that was assigned to the man whose job it was to bring refreshments to the crew—we didn't want him dropping our rapidly diminishing beer supply into the depths of the swamp as he balanced along a log. The rest of us took our chances, and from time to time there'd be a loud splash followed by even louder cursing as someone slipped off a log and fell in.

As I mentioned earlier, landing pile fires would sometimes burn underground over the winter and still be going late the following spring. We had a couple that burned under the dirt for two years before they were finally put out. Every spring I'd go to all the piles that had been lit the previous fall to determine if they were completely out. I had very effective ways of detecting underground heat, including walking through the ash barefoot to check for hot coals. This method never did sit well with the odd forest officer who was present when I used it, as the Forest Service preferred to use infrared scanning techniques with a handheld device called a Probeye. I'd had a chance to examine one of these gadgets while attending a fire management seminar. There weren't any lighted cigarettes in the room, so I had scanned the few females present as a test (surreptitiously, of course, so I wasn't accused of being a chauvinist pig).

The Probeye was a very sensitive device, which sometimes gave rise to false readings, as was the case on an escape up Bush River when the Forest Service kept insisting that there was a hot spot at one end of the fire. Everything was dead cold, and we just couldn't locate the spot they would find each morning when they flew over. Finally I figured out that the "hot spot" was actually a nest of birds in the top of a snag. We left them in peace and moved on.

Putting out underground fires with water was slow, dirty work, particularly if you encountered a really hot spot as you were standing there holding onto the nozzle. There'd be a steam explosion that would blast hot ash and other material into your face. If the temperature was below freezing, as it was when we had to put out a fire that was too close to the dewatering site at Bush Harbour, the spray drifting back with the wind would form a layer of ice on your clothes. (Fortunately that day a nearby loader operator took pity on us and brought over a half-full bottle of whisky.) Standing there holding the hose wasn't too bad on a really hot day, however, as you'd get cooled off by the spray. It got more interesting if there happened to be tree planters working nearby, as the female planters often worked topless. We tried not to stare at them, but our presence didn't seem to bother them; in fact, they came over and sat with us at lunchtime now and then.

One year a tree-planting crew set up a sizeable camp right beside the main logging road on the West Columbia, complete with a makeshift shower. This facility was right out in the open, which led to traffic slowdowns whenever any of the women were using it, as each logging truck crawled past in low gear with the driver peering out his window. This would be the only time the dust on the road would be reduced to the point where there was (in)decent visibility. You had to be careful though; one day a truck turning a corner near the camp encountered one of the male tree planters strolling down the middle of the road stark naked, playing a guitar.

•

It was always rather sad when the crew did their last day of pile burning before they were laid off until the following year. The job would become a bit boring without them around to liven things up. There was always time for one last bit of humour, though. After lighting the last pile of the season one year out in the Beaverfoot, one of the crew was about to drive the crewcab clear of the landing. As the truck started to move, I dropped the hose from the tank of slashburning fuel over the tailgate and flipped on the pump switch. The driver took off, unaware that there was a stream of fuel being pumped onto the dirt road behind the truck. After he'd gone a short distance, I dropped a lit match onto the fuel, and as I watched the flame chase after the truck, I called him on the radio, telling him to look in his rear-view mirror. He couldn't stop to shut off the pump, as the flame would have caught up with the truck and set it on fire, since the back of it was soaked in spilled fuel. He had to keep driving until the fuel tank pumped itself dry. When he eventually returned, he was a bit upset, and he asked, "What if the truck had stalled?" I replied that it didn't matter, as we'd finished pile burning for the year and therefore had no further need for either him or the truck.

After the slashburning crew had been laid off for the season, I'd continue to burn a few piles on my own. Now and then I'd get help from others on the Woodlands staff, two of whom came along with me to burn some near Invermere one wintry morning. Due to poor coordination one threw a match just as the other was throwing more fuel on the same pile. The fuel he threw caught fire and he panicked, turning toward his partner, who got hit with the flaming gasoline-diesel mix. I heard shouts, and I turned around in time to see both of them on fire, rolling in the snow. Fortunately both were wearing rain gear, which saved them from getting burned. I was unable to go to their aid, as I was rolling on the ground laughing, because one of them was a volunteer firefighter and should have known to be more careful. However, I wasn't laughing later on that day when we were cutting off trees that had fallen onto a waterline ditch. There was a layer of ice that seemed thick enough to walk on—but I soon discovered it wasn't, and I fell through, my boots filling with freezing water. I was so mad I threw the chainsaw

I was using through the ice as well. It was running when it went through, which didn't do the engine much good. Fortunately it happened to be my supervisor's saw.

●

"It never got weird enough for me."
—Hunter S. Thompson

EPILOGUE

I was fortunate to have done all my burning back in the days before the outdoors was crawling with ecotourists and mountain bikers. We didn't have to worry about spoiling their view of the mountains with smoke, or the amount of carbon dioxide we were releasing into the atmosphere, as the greenhouse effect wasn't an issue at the time. Not that we'd have let a little thing like global warming stop us from lighting up. We weren't the only ones making smoke in that part of the country anyway—the CPR would regularly burn piles of old railway ties as part of their track maintenance work. The line of fires dotted along the mountainside right-of-way looked quite pretty at night.

By the fall of 1989 my years of working on steep terrain were taking their toll, and an old joint injury was making each trip up the mountainside increasingly difficult. This led me to finally give up slashburning and move back to the coast, where I ended up getting back into forest fire suppression, this time with Kusawa Contracting, a company that provided their services to the BC Forest Service. Kusawa had been started by a couple of ex-IFFS employees, so I felt very much at home. Eventually this work led to a position with the BC Forest Service, and I actually became a forest officer myself, which I never would have imagined possible.

As part of my duties I would be sent out to issue burning permits to property owners who wanted to burn debris piles, which was quite a switch. After having been on the receiving end of these official bits of paper, I was very sympathetic to the people I wrote them out for, and I managed to find the odd loophole in the regulations that would make life a bit easier for them. I even ended up issuing a few of the unpopular 242 instructions, and again I'd try to make them as painless as possible while staying within the regulations of the Forest Act.

By an odd twist of fate, when I was sent over to assist on a wildfire outside Bella Coola, I met up with a forest officer I'd had arguments with some fifteen years earlier. It turned out the fire was the result of an escaped slash-burn that had been conducted by none other than my old adversary, who'd left the Forest Service and was now a private contractor. When he saw me walking up the road, he stared at me in shock.

"It's you!"

"Yes, it is, and now the boot's on the other foot," I replied with satisfaction.

We got on fine after that, and I ended up having a beer or two with him a few days later. I'm glad I had the opportunity to resolve things—sadly, he's no longer with us.

After a few years I got out of the fire business altogether when I started working in the logging road deactivation program run at the time by Forest Renewal BC. There was a certain amount of irony to it, as I was now involved with construction of water bars on unused roads. These features are designed to shed water off the road surface at regular intervals in order to reduce erosion, and whenever I'd been requested to put them in on fireguards back when I was slashburning, I'd always flat-out refuse. What goes around comes around, I suppose.

The only burning I get to do now is the occasional pile of branches in the backyard. I still have my old burning jacket, which I wear on these occasions. It's long beyond the point of repair and should really be thrown on the fire as well, but I just can't bring myself to do it—it's a link to the past, albeit a grubby one. From what I understand, there's virtually no slashburning being done anymore by the forest products company I worked for back in the Golden days. The smoke has long since dispersed, but the memories remain, and I often think about the crew who helped make it unforgettable:

Pat · Howie · Grizz · Bass · Paul · Jacques
Madgrass · Len · Brad · Linda · Rocky · Jim
Rip · Brian · Jack · Liz · Tom · Fabian · Burt
Mike · Glen · Grant · Charlie · Sven · Lars
Val · Ken · Ed · Dood · Dwayne · Pete · Walter
and a few others whose names I've forgotten.

GLOSSARY

air tanker: An aircraft fitted with holding tank(s) and capable of delivering water to forest fires. Also known as a water bomber.

Amex: An ammonium nitrate–based explosive, also called ANFO (*a*mmonium *n*itrate and *f*uel *o*il).

bear paw: A piece of plywood attached to a helicopter skid, designed to prevent it from sinking when landing on soft ground or snow.

bird-dog officer: A forest officer who rides in the plane that accompanies an air tanker and is responsible for determining air-attack strategy.

block: see **cutblock**.

boneyard: A small area cleared down to mineral soil, where burned wood debris is placed after being extinguished.

broadcast burning: The application of fire to a logged-over land area to prepare ground for replanting and/or to reduce fire hazard.

burn boss: The individual in charge of a slashburning operation. If it gets away, he automatically becomes a fire boss.

cap box: A small wooden box with a padded interior used for carrying blasting caps.

Cat: Short for Caterpillar, the company name, and used as a generic term for a bulldozer.

catguard: A fireguard constructed by a bulldozer.

choker: A length of steel cable used for attaching logs to a bulldozer, skidder or high-lead rigging during logging operations.

CIL: Canadian Industries Limited, a manufacturer of explosives.

cold trailing: The process of hand-testing to determine if there are remaining hot spots on a forest fire or controlled burn.

come-along: A hand-operated portable winch used to provide pulling power, often containing a steel cable of unknown breaking strength.

convection burn: An ignition method that uses concentric strip firing, moving outwards from a central point to provide indraft.

convection column: A rising column of smoke that has significant vertical size and is produced by a forest fire or broadcast burn.

conventional: The method of tree harvesting that uses ground-based equipment such as bulldozers and skidders to extract the logs.

crewcab: A type of four-door pickup truck that has a longer wheelbase and extra seating for crew transport.

crown: The upper part of a tree that includes branches and needles.

crowning: The term applied to a situation in which upper parts of one or more trees are on fire at the same time. A cause for concern.

crummy (or crummy bus): A vehicle, larger than a crewcab, used for transporting logging crew. The name generally describes the condition of its interior.

cutbank: The dirt or gravel bank on the high side of a logging road.

cutblock (or block): An area of land in a forest within which logging is to take place or has taken place.

cyclic: A helicopter control that enables the pilot to change the main rotor blade angle.

dewatering: Removing log bundles from the water after they've been towed down the lake by tug.

drafting: Filling a tanker truck from a water source such as a creek.

driptorch: A handheld canister with a pipe extension that drips a diesel-gasoline fuel mix over a burning wick when inverted.

duff: Partly decomposed organic material on the forest floor. Otherwise known as humus.

E cord: A type of detonating cord used for connecting explosive charges together.

escape: A condition where slashburn has escaped the planned boundary into an adjacent timber stand.

faller: A key member of a logging crew who does the tree cutting.

fairlead: A winch attachment that uses an arrangement of steel rollers to guide cable onto the drum.

fire boss: The individual in charge of control operations on a forest fire.

Firecord: A type of detonating cord that was specifically designed for blasting fireguard on forest fires.

fireguard: A combined cleared trail and trench dug to mineral soil, constructed for the purpose of controlling a fire. Can be built by hand or bulldozer.

fireline: The control line from which a fire is fought. Can be a constructed fireguard or a natural firebreak such as a road or a creek.

fire pack: A firefighting tool kit for use in forest fire initial attack, which also contains canned food and drinking water of varying quantity and quality.

fly rock: Rock fragments propelled through the air from a blast site by the force of an explosion.

FMC: Short for Food Machinery Corporation, a brand of tracked log skidder.

fringe damage: Minimal damage to the trees just outside the burn boundary.

gorse: A spiky inflammable plant that was introduced to New Zealand and other parts of the world by homesick Scotsmen.

gravity system: A water delivery system used on a hillside that utilizes the force of gravity to provide adequate pressure at the business end of the hose.
hand guard: A fireguard constructed using hand tools, with the expenditure of much physical labour.
hangover: An underground fire that has persisted through the winter from the previous burning season. Also applied to the unpleasant sensation persisting from the previous night's drinking session.
hazard sticks: A set of standardized wooden dowels used for determining the moisture content of forest fuels.
helipad: A structure built for a helicopter to set down on. Can range in quality from a couple of logs set on the ground to a crib constructed of interlocking logs.
helispot: A small clearing in the forest, either a natural opening or a man-made clearing, that's used for a helicopter landing site.
helitorch: An ignition device slung beneath a helicopter that's used for lighting slash on broadcast burning operations.
high-lead: The method of extracting logs from a cutblock using a cable yarding system that lifts them partially or completely off the ground.
indraft: Air drawn into a broadcast burn.
Interagency Mutual Aid Agreement: A reciprocal agreement between the US states and Canadian provinces whereby they lend each other forest fire suppression crews.

Jackall: A tool used for raising vehicles in order to change flat tires.

Jake brake: A generic name for an engine device installed on some large diesel truck engines that uses compression release for slowing the vehicle.

landing: The level area on a cutblock where logs are loaded onto logging trucks. Also applied to a pile of waste material (branches, tops, rotten logs, etc.) that has been pushed up on the site.

light-up: The start of the slash ignition sequence.

line locator: The individual who marks the route for fireguard construction with flagging tape.

lowbed: A heavy truck-and-trailer combination used for transporting logging equipment such as bulldozers.

magazine: A concrete shed or steel box used for storing explosives.

main line: The heavy steel winch cable on a bulldozer.

manuka: A native New Zealand plant that is somewhat less inflammable than gorse.

money clouds: An approaching thunderstorm that may lead to forest fire overtime.

monsoon bucket: A water container slung below a helicopter, used for dropping water on fires.

mop-up: The process of extinguishing every last hot spot on a forest fire or slashburn.

Mountain Red: A brand of cheap red wine that, despite its name, is not recommended for consumption at elevations above five thousand feet during working hours.

napalm: A gelled gasoline that gained notoriety through military use in the Vietnam War.

oiler: An engine room assistant.

OSHA: Short for the Occupational Safety and Health Administration, the US equivalent of the WCB.

overachievement: The result of a slashburn that's a little more active than expected. See also **escape, fringe damage**.

overtime: The part of the day most looked forward to by firefighters.

piss can: A water container with a spray pump attached that's carried on the back and used for extinguishing small fires. Its official designation is "hand tank pump."

pony winch: The secondary winch on bulldozers used for logging operations.

Primacord: A type of detonating cord; regularly used as another name for detonating cord in general.

Primaflex: A powerful type of detonating cord.

Probeye: An infrared thermal viewer used to detect hot spots during mop-up operations on forest fires and slashburns.

Pulaski: A firefighting tool that is a combination axe and hoe.

relay tank: A portable water tank used to hold water during pumping operations that require the intermediate storage of water as it is being sent uphill.

repeater: A mountaintop radio installation used for extending communication range.

ripper: A heavy steel claw or claws mounted at the rear of a bulldozer, used for loosening densely compacted materials during road construction.

ROPS: Short for "rollover protective structure," a strengthened steel canopy on heavy equipment such as bulldozers and skidders, designed to protect the operator in the event of a rollover.

sector boss: A position in the forest fire control organizational hierarchy.

Shindaiwa: A small portable water pump with a two-stroke engine.

silviculture: The science of tree cultivation.

skid: The part of the helicopter landing gear that contacts the ground.

skidder: A rubber-tired machine used for hauling cut trees and logs to the landing.

skid trail: A dirt trail constructed for the purpose of dragging logs by Cat or skidder to the landing.

Sky Spider: The name given to rappel crewmen back when the program was being run by International Forest Fire Systems Ltd. (IFFS).

slash: Debris remaining on the ground after the completion of logging operations. Does not include discarded cable, broken machinery parts or empty liquor bottles.

slip-on tank unit: A water tank with attached fire pump, designed to fit in the back of a pickup truck.

snag: A standing dead tree.

spot burning: Lighting only accumulations of slash within a block rather than the entire block.

spot fire: A small fire outside a fireguard, usually caused by wind-thrown embers.

spotter: The individual responsible for deployment of rappel crewmen onto a forest fire from a hovering helicopter.

stringer: A log acting as a support beam for a wooden bridge.

strip ignition: The method of broadcast burn ignition used on hillsides where the slash is lit in sequential parallel strips from the top down.

sump: A hole or depression in the ground used for water storage.

swamper: A bulldozer operator's assistant, who performs helpful tasks such as dragging out the winch cable should the machine get stuck in soft ground.

Thermalite connector: A small metal cylinder crimped onto the end of blasting cap–fuse assembly that contains a substance that's easily lit with a match or lighter.

UIC: The Unemployment Insurance Commission, the federal government agency that subsidized many slashburners' winter "inactivities." Now referred to as EI (Employment Insurance).

Ukrainian Water Bomber: A log skidder with a water tank attached, used for hauling water during firefighting operations on steep terrain.

Wajax: A portable water pump with a two-stroke engine, most commonly used on forest fires.

WCB: The Workers' Compensation Board (now WorkSafe BC).

windrowing: The process of piling slash in long rows.

yarder: A stationary logging machine that uses a cable system to pull logs from where the trees have been felled to the landing.

ACKNOWLEDGEMENTS

The fact this book exists is due in large part to encouragement from my wife, Pam, and brother, Adrian. The former listened patiently as I related some of the incidents that had taken place on the fireline. The latter provided invaluable help with the photographs and re-drew my crude map.

I'm indebted to my former supervisors at Evans Forest Products for tactfully ignoring rumours of strange events taking place on my burning operations. They, along with the rest of the Woodlands staff at Evans, were very tolerant of the fact I didn't always do things in a conventional way. I can't recall ever being reprimanded for pushing things a bit too far, although in retrospect there was more than one occasion when I should have been.

I'm further indebted to the forest firefighters of previous generations who developed tools and techniques for controlling wildfire. Credit is also due to those who came up with the ways and means of using fire scientifically as a forest management tool.

I consider myself fortunate to have worked in a part of British Columbia that was relatively unspoiled, at a time when there were fewer restrictions on how the work was carried out. I was privileged to have met and worked with many fine individuals during the years I spent in the woods: loggers, equipment operators, helicopter pilots, foresters, camp managers, camp cooks and of course, the slashburners.

Thanks also to Pam Robertson for editing my manuscript, and Brianna Cerkiewicz, Nicola Goshulak and other editorial staff at Harbour Publishing for their work on converting the manuscript into this book. And finally, thanks to Anna Comfort O'Keeffe for accepting the submission.

ABOUT THE AUTHOR

Nick Raeside was seven years old when he fought his first fire—barefoot—in New Zealand. After upgrading to sandals, learning to make napalm on the kitchen stove and moving to Canada, he headed straight for the woods. He worked a variety of jobs including forest firefighter, land surveyor's chainman, forestry official and, of course, slash-burner. This is his first book. He lives in Nanoose Bay, BC.